Fundamentals of
CYCLIC STRESS AND STRAIN

Fundamentals of

CYCLIC STRESS AND STRAIN

Bela I. Sandor

The University of Wisconsin Press

Published 1972
The University of Wisconsin Press
Box 1379, Madison, Wisconsin 53701

The University of Wisconsin Press, Ltd.
70 Great Russell St., London

First printing

Printed in the United States of America
For LC CIP information see the colophon

ISBN 0-299-06100-0

To JoDean Morrow, teacher and friend

Contents

Illustrations ix

Symbols and Abbreviations xv

Introduction 3

1 Static Behavior 5

1.1 Static and Quasi-static Loading 5
1.2 Fracture and Toughness 14

2 Cyclic Loading 17

2.1 Cyclic-Loading Conditions 17
2.2 Cycle-Dependent Material Response 26
2.3 Hysteresis Loops with Cycle-Dependent Response 28
2.4 The Qualitative Prediction of Material Response 31
2.5 Cycle-Dependent Changes in Mechanical Properties 32

3 The Fatigue Life 39

3.1 Analysis of the Strain-Life Curve 42
3.2 The Transition Fatigue Life 51
3.3 The Desirable Life Curve 52

4 Energy Considerations in Fatigue 54

4.1 Failure Prediction 55
4.2 Examples 61

5 Cumulative Damage 64

5.1 The Exhaustion of Life 68
5.2 Plasticity and Cumulative Damage 71
5.3 Examples 82

6 Mean Stresses and Mean Strains 85

6.1 Fatigue Life with Mean Strains 90
6.2 Fatigue Life with Mean Stresses 92
6.3 Fatigue with Mean Stress and Mean Strain 97
6.4 Theory of Fatigue Damage with Mean Values 100
6.5 Residual Stresses 103
6.6 Relaxation of Mean Stresses 106

7 The Notch Problem 113

7.1 Fatigue Prediction of Notched Members 116

7.2 The Control Condition at a Notch 120
7.3 Notches and Residual Stresses 123
7.4 Single and Multiple Overloads 126

8 Unusual Fatigue Behaviors 131

8.1 The Shape of the Hysteresis Loop 132
8.2 Cyclic Creep without a Mean Stress 133
8.3 Hardening and Softening under Constant-Amplitude Cycling 134
8.4 Strengthening by Strain-Aging 135
8.5 Life Reduction by Electropolishing 136
8.6 Size Effect in Fatigue 137
8.7 Crack Growth in Compression 137
8.8 Simultaneous Creep and Cyclic Deformation 138
8.9 Fatigue of Electroplated Parts 138
8.10 Fatigue of Atomic-Chain Fiber 139

9 The Approaches to Fatigue Problems 141

9.1 Post-mortem Examinations 141
9.2 Fatigue-Damage Prediction and Indication 145
9.3 Material Selection 146
9.4 Fatigue-Damage Alleviation 151

Conclusions 155

References 159

Index 165

Illustrations

FIGURES

1 Example of cyclic loading, airplane 3
2 Example of cyclic loading, bridge 3
3 Example of cyclic loading, ship 3
4 Model of axially loaded specimen 6
5 Monotonic stress-strain curve 7
6 Large deformation followed by unloading 7
7 Elastic recovery after various deformations 9
8 Example of overstraining necessary to produce a permanent deformation 9
9 Model of the plastic flow and elastic recovery processes 10
10 Model of complex flow and recovery processes 11
11 Grain structure of polycrystalline material 11
12 Compressive loading after large tensile deformation 12
13 Sign convention for stresses and strains 12
14 Quadrants of the stress-strain plane 13
15 One way of loading into second quadrant 13
16 Loading into fourth quadrant from compression 13
17 Loading into fourth quadrant from tension 13
18 Strain after fracture and definition of toughness 14
19 Stress vs. plastic strain to fracture 15
20 Desirable combination of strength and ductility 16
21 Monotonic strength and ductility attainable in several metals 16
22 Schematic of cyclic loading 18
23 Definition of cyclic stresses 18
24 Definition of cyclic strains 18
25 Phase relation of cyclic stresses and strains for elastic behavior 19
26 Cyclic stress-strain plot for elastic behavior 19
27 Idealized tensile and compressive stress-strain curve 20
28 Development of the hysteresis loop 20
29 A complete hysteresis loop 21
30 Model of cyclically reversed plastic deformations 21
31 Definition of quantities associated with the hysteresis loop 22

32 The additive nature of hysteresis energy 23
33 Laboratory specimen with extensometer for longitudinal strain measurement 25
34 Laboratory specimen with diametral clip-gage extensometer 25
35 Cycle-dependent material responses under stress control 26
36 Cycle-dependent material responses under strain control 27
37 Stress vs. strain for stress control and cycle-dependent hardening 28
38 Stress vs. strain for stress control and cycle-dependent softening 29
39 Example of hysteresis loops for stress control and cyclic softening 29
40 Stress vs. strain for strain control and cycle-dependent hardening 30
41 Stress vs. strain for strain control and cycle-dependent softening 30
42 The role of initial material condition in cyclic stress-strain response 31
43 Cyclic response of mild steel with respect to the load range 32
44 Cycle-dependent changes in stress-strain response 32
45 Examples of cyclic stress-strain curves 33
46 Influence of hardness on the cyclic stress-strain response of SAE 4142 steel 34
47 One definition of the cyclic stress-strain curve 35
48 Experimental determination of the cyclic strain-hardening exponent 36
49 Gradual change from monotonic to cyclic stress-strain curve 36
50 Uncertainty of stabilization in stress-strain response 38
51 Stress vs. life, or *S-N*, curve 39
52 Rates of fatigue-crack growth 40
53 Strain vs. life curve 41
54 Test equipment for the control and measurement of cyclic stresses, strains, or deflections 42
55 Example of the cyclic-strain components 43
56 Superposition of the strain components 43
57 Fatigue properties of SAE 4340 steel 44
58 Fatigue properties of 7075-T6 aluminum alloy 45
59 Fatigue properties of titanium alloy Ti-8Al-1Mo-1V 46
60 Correlation between the fatigue ductility coefficient and the monotonic fracture ductility 47

61 Exhaustion of ductility in tension 48
62 Model for exhaustion of ductility 48
63 Simple schemes of cyclic plastic strain vs. life 49
64 Transition fatigue life as a function of hardness for steels 51
65 Practical considerations concerning the most desirable life curve 52
66 Fatigue toughness as a function of life 55
67 Model of plastically deformed zones at crack tips 56
68 Analysis of cyclic work 58
69 Random loading 64
70 Simple step loading, showing high-to-low and low-to-high sequences 65
71 Multiple-step programming 65
72 Schematic of cycle-dependent adjustment of independent variable under step change of control function 66
73 Stress adjustments in copper under controlled cyclic strains 67
74 Damage accumulation in high-to-low step loading 68
75 Damage accumulation in low-to-high step loading 69
76 Model of linear damage rates 72
77 Piecewise linear representation of damage under complex cyclic plastic straining 74
78 Nonrecoverable strains at small stresses 76
79 Stress vs. life for mild steel 77
80 Strain vs. life for mild steel 77
81 Rates of change of cyclic stresses in copper 79
82 Schematic functions of cycle-dependent plastic strains 80
83 The generation of mean stresses and mean strains 86
84 Sole ways of returning to zero stress and strain after plastic deformation 87
85 The dependence of mean stress on prior plastic strain and on applied stress range 88
86 The incompatibility of large mean stress and large cyclic stress 88
87 Elastic behavior with mean stresses and mean strains 89
88 The incompatibility of large mean stress and large cyclic plastic strain 90
89 Schematic of increasing mean strain 90
90 The effect of prestrain on fatigue life 92
91 Schematic of fatigue lives with mean stresses 93
92 Examples of fatigue lives under tensile mean stresses 93
93 The idealized effect of mean stresses 94
94 Three approaches to design with mean stresses 95
95 A simple analytical method of accounting for mean stresses 96

96 Fracture surfaces of mild steel tested in axial push-pull 97
97 Model of cyclic creep caused by a mean stress 98
98 Model of cycle-dependent buckling 98
99 Deceleration of cyclic creep 99
100 Model of changing crack shape in fatigue 101
101 Stresses and strains at different locations in a cracked member 102
102 Model of surface peening 104
103 Residual stress distribution; cyclic stress is superimposed on local residual stress 105
104 Relaxation of a mean stress in SAE 1045 steel 107
105 The exponential nature of mean stress relaxation 107
106 Schematic of mean stress relaxation for a cyclically softening material 108
107 Schematic of mean stress relaxation for a cyclically hardening material 109
108 The role of cyclic plasticity in mean stress relaxation 109
109 Schematic of mean stresses at half-life under strain control 110
110 Stress distribution in a notched member 114
111 True stress concentration 116
112 Prediction of fatigue life of notched members 119
113 Model for the behavior of notched members 121
114 Residual stresses created by an overload 124
115 Analysis of stresses in an overloaded notched member 125
116 The effect of tensile and compressive overloads 126
117 Transition from large to small loads without leaving significant residual stresses 127
118 Residual stress after multiple overloads 127
119 Early method of preflight testing 128
120 Wing before and after test loading 128
121 Wing in flight 129
122 The effect of landing, at some critical points of an airplane 129
123 Distorted hysteresis loop caused by stress relaxation 132
124 Bendback in hysteresis loop of cast iron 132
125 Cyclic creep of cast iron under fully reversed stresses 133
126 Idealized curves for the relaxation of mean stresses 134
127 Complex instability in stress-strain response 135
128 Strengthening by strain aging in low-carbon steels 135
129 Fatigue crack growth under compressive loading 137
130 Cyclic loading of hypothetical atomic-chain specimen 140
131 Fatigue fracture surface appearance 142
132 Characteristics of spectrum loading 143

133 Improvement of fatigue life by reduction of stress concentration 144
134 The effect of beam thickness on life under cyclic load control 144
135 The effect of beam thickness on life under deflection control 145
136 Schematic representation of cyclic strain resistances 150
137 The desired segregation of strength and ductility at the minimum section 151
138 Influence of grain size and cold work on the flexural fatigue strength of 70–30 brass 152
139 Fatigue damage alleviation by remachining 153

TABLES

1 Cycle-dependent changes in mechanical properties 37
2 Strain-controlled step tests on 2024-T4 aluminum alloy; no mean stresses 71
3 Strain-controlled step tests on 2024-T4 aluminum alloy; mean stresses 111
4 Stress-concentration factors for various alloys in rotating bending 115

Symbols and Abbreviations

A	area
b	fatigue-strength exponent, Basquin's exponent
BHN	Brinell Hardness Number
c	fatigue-ductility exponent
d	diameter
$d\epsilon_p$	infinitesimal plastic deformation
D	damage number, a constant
Δ	deflection
ΔD	fraction of damage number D
ΔN_i	number of cycles in a block at the ith level of loading
ΔL	deformation
ΔW	energy per cycle
$\Delta\epsilon$	range of strain from peak to peak
$\Delta\epsilon/2$	strain amplitude
$\Delta\sigma$	range of stress from peak to peak
$\Delta\sigma/2$	stress amplitude
e	engineering strain, change in length per original length; also, exponential base
E	modulus of elasticity
ϵ	true strain, logarithm of (ratio) instantaneous length per original length
ϵ_a	total-strain amplitude, half of peak-to-peak range
ϵ_e	elastic strain
ϵ_f	true fracture ductility, plastic strain after monotonic fracture
ϵ_f'	fatigue-ductility coefficient
ϵ_{ft}	total strain at instant preceding fracture
ϵ_m	mean strain, average of peak strains
ϵ_n	nominal total strain
ϵ_p	plastic strain
F	force
γ	surface energy
K	strength coefficient, in units of stress; occasionally, a constant without units
ksi	thousand psi
K_ϵ	strain-concentration factor

K_f fatigue-strength reduction factor
K_p plastic-strain concentration factor
K_σ true stress-concentration factor
K_t theoretical stress-concentration factor
L length
m a constant
n strain-hardening exponent for static loading
n' strain-hardening exponent for cyclic loading
n_i number of cycles at the ith level of loading
N number of cycles
N_f cycles to failure
N_{fi} fatigue life at the ith level of loading
N_T transition fatigue life (elastic strain = plastic strain)
P force
psi pounds per square inch
q notch-sensitivity factor
R plastic-zone size; also, stress ratio = $\sigma_{min}/\sigma_{max}$
S engineering stress, force per original area = P/A_o
σ true stress, force per instantaneous area = P/A_i
σ_a stress amplitude, half of peak-to-peak range
σ_f true fracture strength in monotonic loading
σ_f' fatigue-strength coefficient
σ_F nominal fracture stress for a notched member
σ_m mean stress, average of peak stresses
σ_n nominal stress, average stress
σ_r range of stress from peak to peak
σ_{ty} tensile yield strength
σ_u ultimate strength
σ_y yield strength
t time
T tensile force
τ shear stress
W_f true toughness in monotonic fracture
W_p fatigue toughness, total plastic strain energy in a given fatigue life

Fundamentals of

CYCLIC STRESS AND STRAIN

Introduction

Fatigue of materials is a common problem that affects everything that moves, and even apparently stationary objects may fail in fatigue if they experience cyclically imposed forces or deformations. There are innumerable problem areas. Automobiles fatigue from rough roads, airplane wings flutter in flight, bridges deflect under each passing vehicle, ships are heaved by the waves, and nuclear reactors are damaged by cyclic temperature changes (Figs. 1, 2, 3).

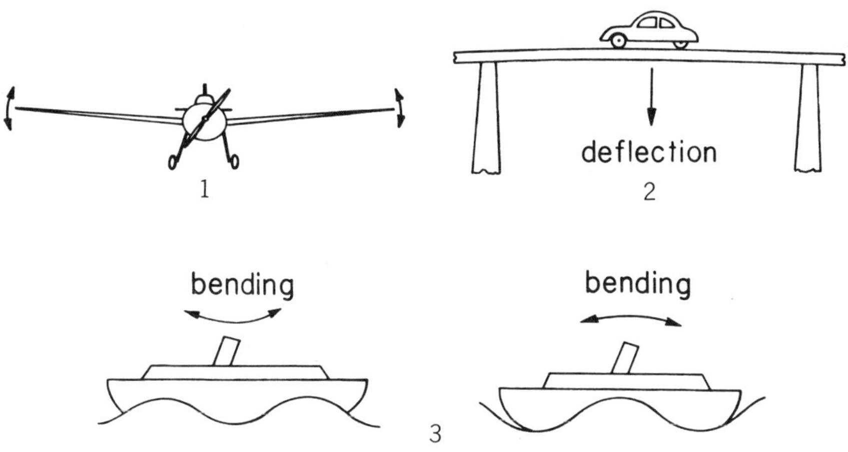

Figs. 1, 2, 3. Examples of cyclic loading

The subject of fatigue is complex and can be studied in many different ways [38].* These range from full-scale structural tests, through laboratory tests of simple-looking specimens, to investigations at the atomic level. From the engineering point of view the approach based on an analysis of stress and strain is one of the most important. The fundamentals of this approach will be discussed in detail in this book.

*Numbers in brackets refer to list of references.

If the reader wishes to remember only one idea from those presented here, it should be the concept that cyclic plastic strain is ultimately responsible for fatigue damage. This will be repeated later, after the subject has been developed, but it cannot be overemphasized. The strength of the concept is derived partially from its ability to unify the apparently different problem areas of low-cycle (short-life) and high-cycle (long-life) fatigue.

Cyclic plasticity is also believed to be the cause of fatigue damage in all materials and under any environmental or cyclic-loading condition. This will not be obvious from the treatment here, which will be limited mainly to discussions of low-cycle fatigue of ductile metals loaded under room temperature condition. The whole matter, however, is one of degree. The view presented will be highly visual, showing the important physical quantities in relatively large magnitudes for display purposes, even when in reality they may be extremely small. Thus low-cycle fatigue concepts can be used at long lives and even for nonductile materials. True fatigue failures in wood, ceramics, polymers, and metals are all believed to be caused by cyclic plastic strains, regardless of any other factors that may be involved in the damage process. Corrosive environments and elevated temperatures, for example, can alter the damage rates substantially by causing such nonfatigue kinds of damage as material loss or creep damage, and by changing the level of cyclic plasticity or tolerance to such deformations. In any case, cyclic loads cause damage under a given set of conditions only to the extent that they result in reversed plastic flows.

The importance of cyclic plasticity is mentioned throughout the text, but it should not obscure the fact that there are other ways of dealing with the fatigue problem. The fracture mechanics approach, for example, provides an extremely useful method for analyzing and predicting crack propagation. Cyclic plasticity may be viewed as a fundamental factor in fatigue damage accumulation, one that plays a role in other approaches as well but a role that may not always be immediately clear to the observer. Much further research and thought will be required to develop a satisfactory overall understanding of cyclic loading and deformations.

1

Static Behavior

1.1 Static and Quasi-static Loading

An effort to understand the fatigue of materials must be based on an extensive knowledge of static stress-strain responses [11, 63]. This must include a knowledge of tensile and compressive behavior, fracture properties, and quasi-static behavior such as that found when unloading from tension and subsequently applying compression.

The basic stress-strain relations are defined with respect to an axially loaded member, as in Fig. 4. The following definitions [59] are relevant to the present discussion:

$$\text{Engineering stress, } S = \frac{P}{A_o} \tag{1}$$

$$\text{Engineering strain, } e = \frac{\Delta L}{L_o} \tag{2}$$

$$\text{True stress, } \sigma = \frac{P}{A_i} \tag{3}$$

$$\text{True strain, } \epsilon = \ln \frac{L_i}{L_o} \tag{4}$$

since $d\epsilon = dL_i/L_i$ where the subscript i refers to instantaneous values. For plastic deformation we may assume constant volume, $A_o L_o = A_i L_i$. Consequently

$$\epsilon = \ln \frac{A_o}{A_i} = 2 \ln \frac{d_o}{d_i} \tag{5}$$

or $L_i = L_o + eL_o$, which leads to

$$\frac{L_i}{L_o} = 1 + e \tag{6}$$

Thus we get

$$\epsilon = \ln (1 + e) \tag{7}$$

and

$$\sigma = S(1 + e) \tag{8}$$

Equation 7 is based on the assumption that the strains are homogeneous (no necking). Equation 8 requires both constant volume and homogeneous strain.

The material's response under gradually increasing load is shown in Fig. 5, where the broken line is the engineering stress-strain curve for a ductile metal. This curve has limited use in understanding the mechanical behavior of materials, and it will not be discussed here. The solid line is the true stress-strain plot for the same metal. The initial linear portion of the curve represents elastic behavior, and there is plastic flow above point A. Plastic flow is inelastic permanent deformation, and it becomes evident when the load is

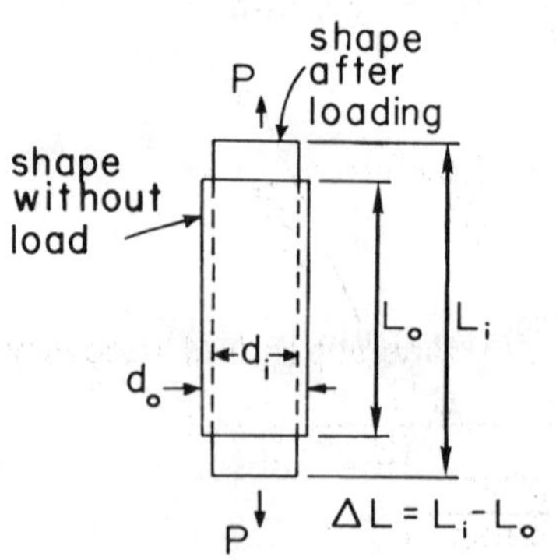

Fig. 4. Model of axially loaded specimen

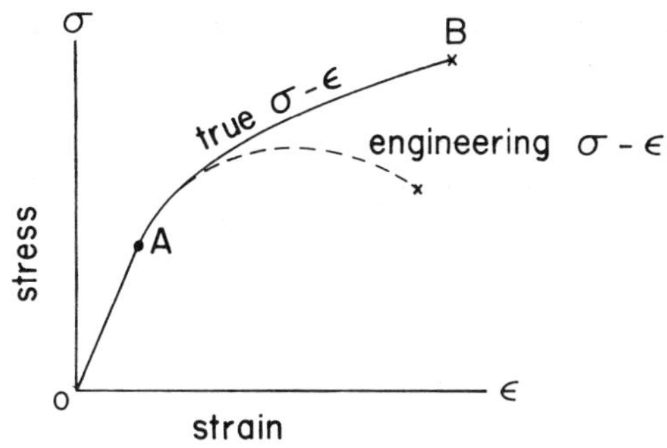

Fig. 5. Monotonic stress-strain curve

removed, as shown in Fig. 6. The unloading curve is along a line parallel to the original elastic line.

In some cases one can also observe anelastic strains which can be recovered in time. Figure 6 shows that immediately after unloading from point *A* the residual strain is at *B*. Allowing sufficient time, the strain relaxes to point *C*. Thus the true plastic strain is the irrecoverable portion $\overline{OC}$. Anelasticity is normally not measured or defined in terms of stress or strain, but in terms of energy, phase angle, or logarithmic decrement [11]. Its significance in cyclic deformation is mainly at very long lives. It will not be discussed further here, since fatigue damage concepts will be based on cyclic plasticity.

The discussion now returns to the true stress-strain curve in Fig. 5. It may be helpful to emphasize that the true σ-ε curve represents true stress vs. true *total* strain. These are the quantities that are most directly measured in a test. The instruments used for measuring strain can only pick up total strain and cannot distinguish the components of this strain without further manipulation.

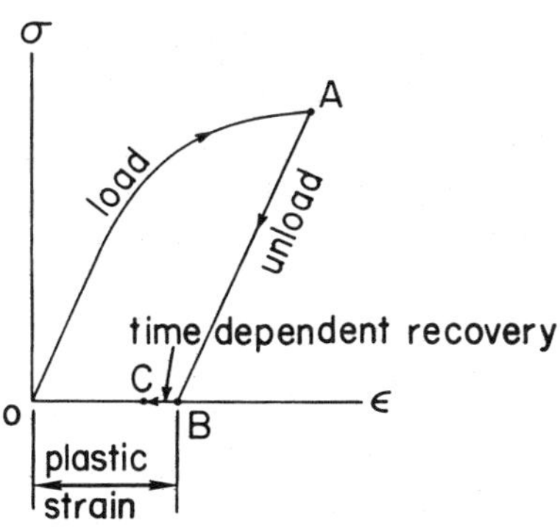

Fig. 6. Large deformation followed by unloading

Mathematically, the components of the true total strain are the following:

$$\text{elastic strain, } \epsilon_e = \frac{\sigma}{E} \tag{9}$$

$$\text{plastic strain, } \epsilon_p = \left(\frac{\sigma}{K}\right)^{1/n} \tag{10}$$

Equation 9 is Hooke's law. Equation 10 is obtained from the power law that relates stress and plastic strain in a strain-hardening material: $\sigma = K\epsilon_p{}^n$ where K is the strength coefficient and n is the strain-hardening exponent. Thus total strain is

$$\epsilon = \epsilon_e + \epsilon_p = \frac{\sigma}{E} + \left(\frac{\sigma}{K}\right)^{1/n} \tag{11}$$

in which ϵ_e is recoverable and ϵ_p not recoverable. According to Eq. 11, every point between O and B in Fig. 5 has these two components of strain. Elastic strains dominate between O and A, and it is very difficult to measure plastic strains in this region. However, plastic strains have been measured in largely elastic materials at stresses as low as a few psi, and one must assume that any finite stress will produce a finite plastic flow in any material (at least in localized zones), even though infinitesimal strains cannot yet be recorded. An appreciation of this concept is essential in learning the fundamentals of fatigue damage, and more will be said about it later.

Another statement of Eq. 11 is that at every point between O and B in Fig. 5 there is immediately recoverable elastic strain present. In fact, the elastic strains gradually increase with increasing stresses up to point B. This is obvious in the linear region, but much less so in the nonlinear portion. The statement is illustrated graphically in Fig. 7. The figure can be interpreted as follows.

When the material is loaded to σ_A, the total strain is ϵ_A. If we unload from σ_A, the material returns essentially to point O. Thus the recovered strain is $\overline{O\epsilon_A}$.

At a load of σ_B the total strain is ϵ_B. Upon unloading from B the material relaxes to the strain at b. The recovered strain is the amount $\overline{b\epsilon_B}$. The strain $\overline{Ob}$ is irrecoverable, and it is the plastic strain caused by actually straining to ϵ_B.

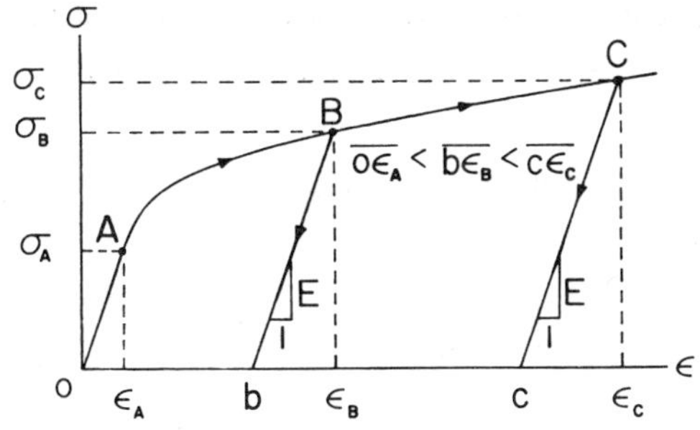

Fig. 7. Elastic recovery after various deformations

The events that follow a loading to σ_c are qualitatively the same as those at point B. A strain equal to $\overline{C\epsilon_c}$ is recovered, and the plastic strain is $\overline{OC}$. Similar statements can be made about any point on the curve.

The three triangles shown in Fig. 7 are geometrically similar and have slopes equal to the elastic modulus. Each vertical leg is the stress at that particular point, and the horizontal leg is the corresponding elastic strain. The importance of this is that if the stress σ and the total strain ϵ are known at any point in the material, the plastic strain ϵ_p can be calculated as follows: since $\epsilon = \epsilon_e + \epsilon_p$,

$$\epsilon_p = \epsilon - \epsilon_e = \epsilon - \frac{\sigma}{E} \tag{12}$$

As an example of the physical meaning of the problem posed in Fig. 7, consider the possible deformations in a spring. Figure 8 shows a spring with an original length L_o, and a length L_1 under load. After relatively small loads one expects the spring to return to its original length L_o, but it is not surprising that large enough loads can cause a permanent deformation (magnitude $= L_2 - L_o$),

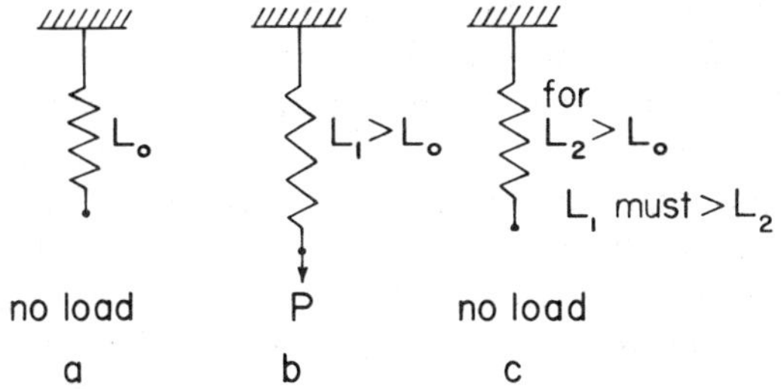

Fig. 8. Example of overstraining necessary to produce a permanent deformation

as in Fig. 8c. The simplest experiments can demonstrate that $L_2 > L_o$ only if the length L_1 (under load) exceeds the final permanently stretched length L_2. In other words, even an overloaded spring appears to have a certain residual resilience.

Perhaps the most difficult aspect of monotonic deformation to understand is that both the elastic and the plastic strains increase with increasing stresses, as shown in Fig. 7. The following crude analogy can be developed to facilitate the understanding of this behavior.

Consider the sawtooth pattern of similar triangles in Fig. 9, and imagine that a smooth curve connecting the tips of the teeth is a stress-strain curve. Starting from point O a ball is pushed up the successively higher ramps. Along each ramp the ball has a certain potential energy above the level O, and much of this energy can be recovered by removing the propelling force P and letting the ball fall back to the ground level. Up to point A, the ball

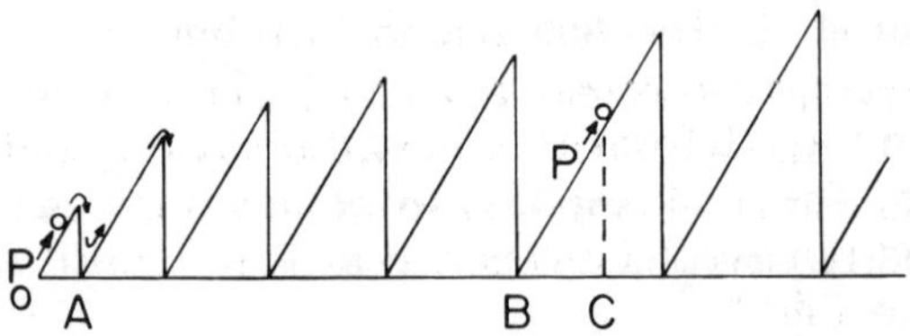

Fig. 9. Model of the plastic flow and elastic recovery processes

could always return to O by its own weight, and this would correspond to the perfectly elastic behavior in the beginning part of the stress-strain curve. The situation past point A becomes more complicated, as only part of the total displacement can be recovered when the propelling force is removed. For example, if the ball is pushed past point B to point C, the total displacement is $\overline{OC}$ while the force P is applied. If P is removed, the ball can fall back to B, the amount of permanent displacement. Working with the peak heights in Fig. 9, an analogy can be made with the curve in Fig. 7. The farther we go from the origin in either of these figures, the greater will become both the recoverable and the irrecoverable displacements.

The above argument can be carried further to deal with the deformation in a real material, which obviously does not behave as discontinuously as the model in Fig. 9. To do this, consider three identical sawtooth patterns taken from Fig. 9 and superimposed on each other with small offsets, as in Fig. 10. The offsets are needed to establish the analogy with a polycrystalline material in

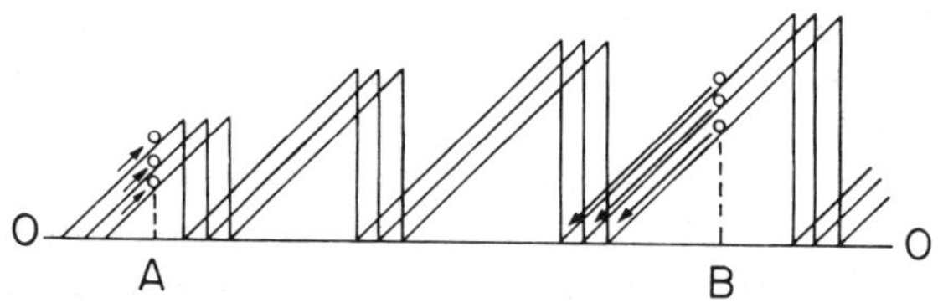

Fig. 10. Model of complex flow and recovery processes

which the grains have a random orientation (Fig. 11) and a correspondingly nonuniform behavior under load. Suppose a ball is pushed on each ramp from left to right in such a way that the three balls are always on the same vertical line while being pushed. This means that the total displacements are the same. Initially, all of these individual displacements are recoverable, as at point *A* in Fig. 10. The situation at *B* in Fig. 10 becomes more interesting, since the lowest ball may be below the highest one at point *A*. Yet on the average, both the elevation and the displacement are larger at *B* than at *A*. The recoverable portions are similarly larger at *A*, on the average. The material in Fig. 11 behaves in the same way. At a given total deformation the local stresses and local strains are not uniform. The total response to external load and the recovery after the load is removed, however, can be described by the earlier discussion about Fig. 7.

Such modeling of the elastic and plastic deformations is naturally simplistic. A more precise analysis of the mechanical behavior requires a familiarity with dislocation and other defect mechanisms. The technical literature is rich in treating these subjects at various levels of complexity, and the interested reader will find references 11 or 63 particularly helpful.

After considering the problems associated with unloading, it is reasonable to go a step further and discuss one or two reversals in the loading. These may be called quasi-static events, as they are between the simple tension or compression test and the full-fledged cyclic loading. In a sense these include every possible stress-

Fig. 11. Grain structure of polycrystalline material

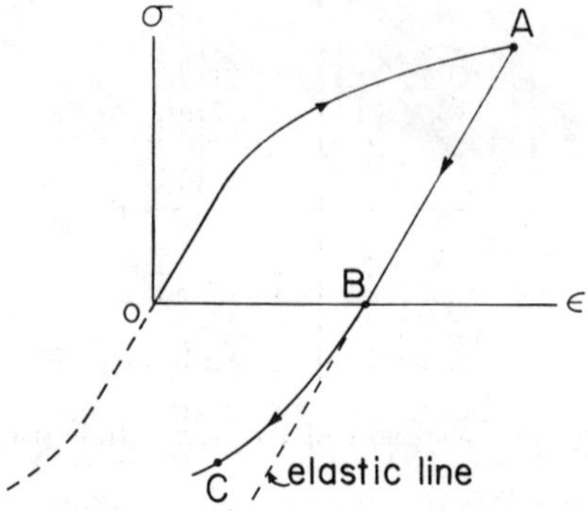

Fig. 12. Compressive loading after large tensile deformation

strain situation that the material can experience before the first full cycle is completed. The quasi-static behavior plays an important role in the understanding of fatigue damage accumulation.

Figure 12 shows one way in which this may occur. The material is loaded in tension to point *A*, then it is unloaded to point *B*, essentially along the elastic line. An immediate reloading in tension would proceed along the same line from *B* to *A*. On the other hand, if at point *B* the load is reversed from tension to compression, the stress-strain curve soon deviates from the elastic line, as the segment *BC* in the figure shows. The early inelastic behavior contrasts with that shown in the first loading from *O* to *A* but is a consequence of the plastic flow before *A* is reached. The phenomenon is described in more detail in discussions of the Bauschinger effect [11, 63].

An interesting and important result of the reversed directions of plastic deformation is the possible combinations of tensile and compressive stresses and strains. Tensile stresses and strains are considered positive and compressive stresses and strains negative. Figure 13 shows the ranges in which the respective signs are applicable. The possible stress-strain combinations are shown in Fig. 14. The significance of these possibilities in fatigue will be discussed later.

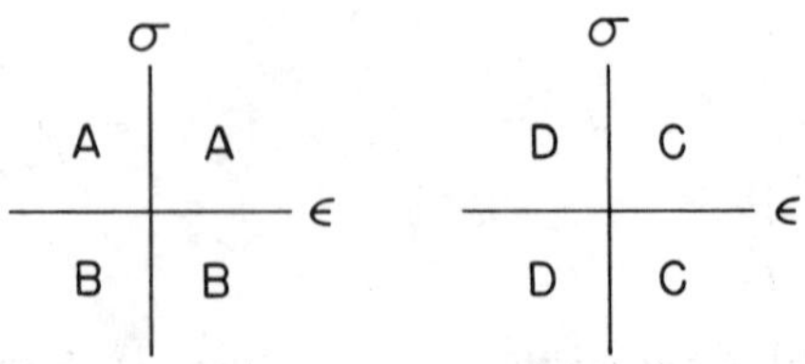

Fig. 13. Sign convention for stresses and strains

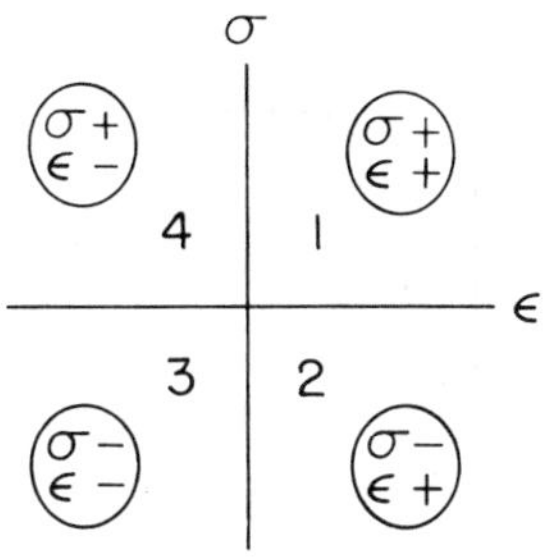

Fig. 14. Quadrants of the stress-strain plane

The combinations of stress-strain signs cannot be arbitrarily produced. A virgin material can be "started" from point 0 into either the first or the third quadrant, but never into the second or the fourth. Each of the even quadrants can be reached in two ways, but only by way of plastic deformations. The second quadrant can be reached by having been in the first quadrant, as in Fig. 12. It can also be reached by going from first to second to third, and back to second, as in Fig. 15, where the unloading from point

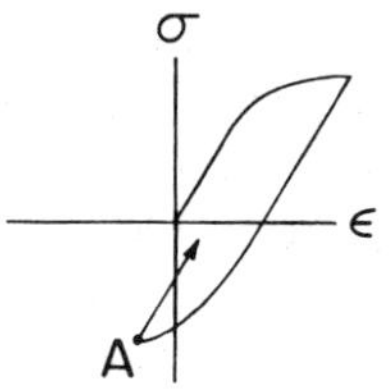

Fig. 15. One way of loading into second quadrant

A is again along the elastic line. Similarly, the fourth quadrant is available only after having been in the third (Fig. 16), or in the first (Fig. 17) if it is a third-to-fourth-to-first quadrant sequence. The generalization is that plastic strain in the first quadrant is necessary to get into the second, and that plastic strain in the third quadrant is necessary to get into the fourth.

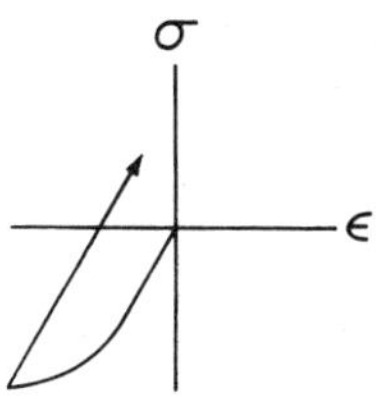

Fig. 16. Loading into fourth quadrant from compression

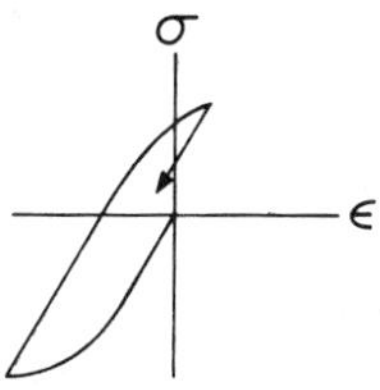

Fig. 17. Loading into fourth quadrant from tension

1.2 Fracture and Toughness

Any material can be loaded to the point of fracture, and any material, even the most brittle, absorbs some energy in the process of being strained to fracture. The information obtained in a monotonic tension test to fracture is important in fatigue analysis and prediction because useful material properties can be found in relatively simple, fast, and cheap ways. In order to prepare for fatigue analysis it is necessary to define and clarify three material constants associated with monotonic fracture.

In Fig. 18, at the instant preceding fracture the true stress is σ_f and the true total strain is ϵ_{ft}. We must refer to total strain at this instant because the material is still holding together, so that all of the strain components are present. When fracture is complete, the stress falls to zero and the strain falls to ϵ_f, the plastic strain after fracture. One can think of this process as unloading from A by separating the load-carrying member. Of course, elastic strain equal to $\epsilon_{ft} - \epsilon_f$ is recovered upon fracture.

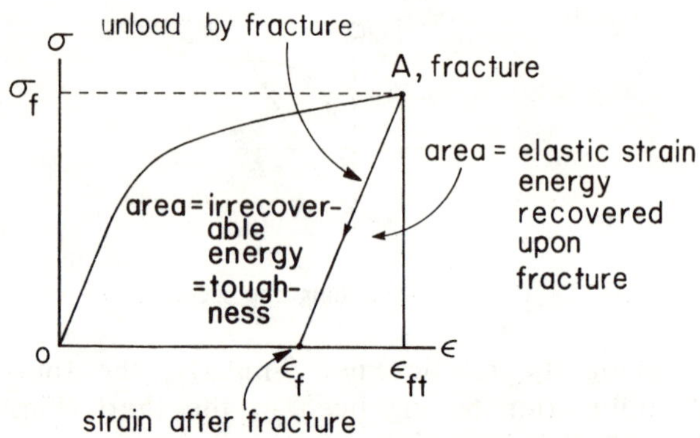

Fig. 18. Strain after fracture and definition of toughness

The quantities σ_f and ϵ_f are called the fracture strength and fracture ductility of the material, and they are assumed constants for a given material. It must be pointed out that ϵ_f is often confused with ϵ_{ft}. The magnitude of the resulting error depends on the fracture strength, the elastic modulus, and the ductility of the material.

An important factor in the resistance of a material to fracture is its toughness, its ability to absorb energy. This can be looked at from both the stress and the plastic strain points of view. The elastic strain energy [4], the area under the linear portion of the stress-strain curve, is high in the case of a steep elastic line and high levels of stress. Unfortunately, even a high value for this energy cannot contribute to toughness, since all of it is recoverable energy. It is through plastic strain that energy is absorbed. The area under the stress-strain curve has the units of energy per unit volume, and thus a tough material is one with a large area under the plastic portion of the curve.

Texts on strength of materials [4] generally define toughness as the area under the engineering stress-strain curve, a practice that cannot provide accurate results. It is much better to use the true stress-strain curve for the determination of toughness, even though it is still possible to introduce large errors by including the recoverable energy. For accurate fatigue damage analysis it is necessary to use the most precise data from monotonic tests. Thus, for the true stress-strain curve of Fig. 18, toughness is less than the total area under the curve. Note that the triangular area above ϵ_f to ϵ_{ft} represents the elastic strain energy in the material just before fracture, and that this energy has been recovered by unloading in fracture. Another way to define the "true" toughness is to plot true stress versus plastic strain directly, as in Fig. 19, and calculate the area under the curve.

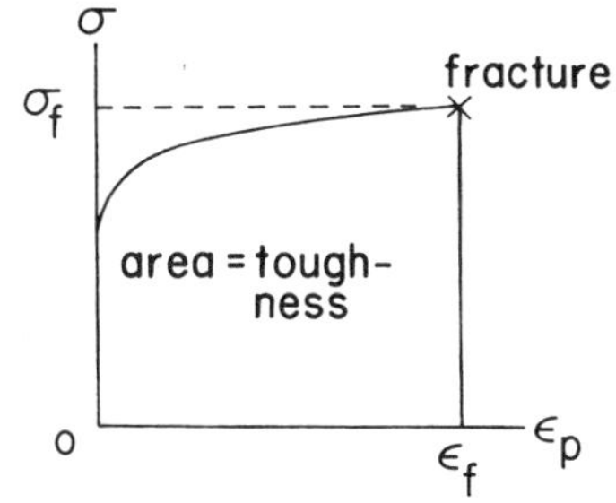

Fig. 19. Stress vs. plastic strain to fracture

Ideally, a material could be highly desirable simply by having a large value of toughness. In practice, it is best to have high toughness as a result of *both* high strength and ductility, as shown in Fig. 20. This is true because, for example, high toughness mainly from high ductility could imply low strength, and consequently low resistance to deformation. There are many practical situations where large deformations, whether elastic or plastic, are not acceptable

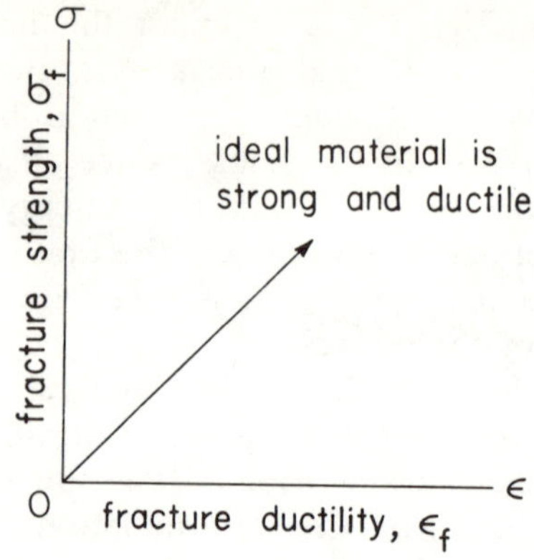

Fig. 20. Desirable combination of strength and ductility

even if fracture is avoided. Figure 21 shows how several metals compare with each other and with the ideal behavior indicated in Fig. 20. Naturally such a diagram is not by itself a sufficient basis for choosing a material for structural applications. Other factors, such as cost and strength per unit weight, should be considered as well.

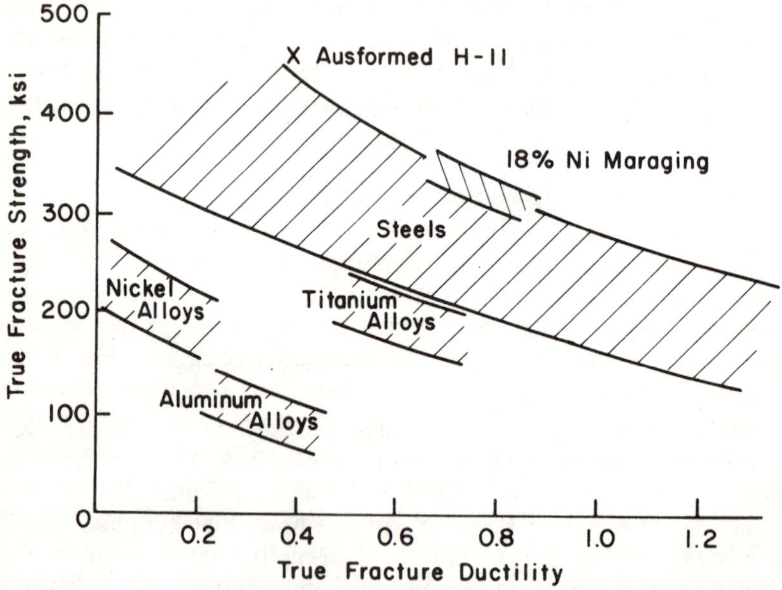

Fig. 21. Monotonic strength and ductility attainable in several metals (from Landgraf, in *High Fatigue Resistance in Metals and Alloys*. Courtesy ASTM)

2

Cyclic Loading

In essence, there are two aspects to the understanding of stress and strain in materials fatigue. First, we must have a complete knowledge of the cyclic-loading conditions imposed on the material. On a large scale this is normally not a difficult task, but in a number of cases, such as those with complex geometries, the problem becomes anything but simple. Second, we should realize that every material has a "personality" of its own and wants to respond to any particular imposed condition in its own way. Satisfactory descriptions of the possible responses are often difficult to come by or have not been available until recently. The generally poor understanding of materials fatigue is largely caused by the lack of knowledge of the material "response" aspect.

2.1 Cyclic-Loading Conditions

The simplest imposed conditions are those in which the loading is periodically and uniformly repeated, as shown in Fig. 22. The loading function may be, in the simple cases, either force, stress, displacement, or strain. In this text stress and strain are used most often, since these show best the material's behavior. For a simple description of the behavior, one of these two quantities is controlled in such a way that the peaks are constant from cycle to cycle, and the other quantity is allowed to vary according to the character of the material. In many common fatigue problems the wave form is not a critical variable, and attention is focused on the upper and lower peaks of the loading. A useful way of thinking of the upper and lower limits is that they are boundary "walls." Regardless of the nature of the control condition (stress or strain), the material is forced to a wall, and then the direction of the forcing is reversed

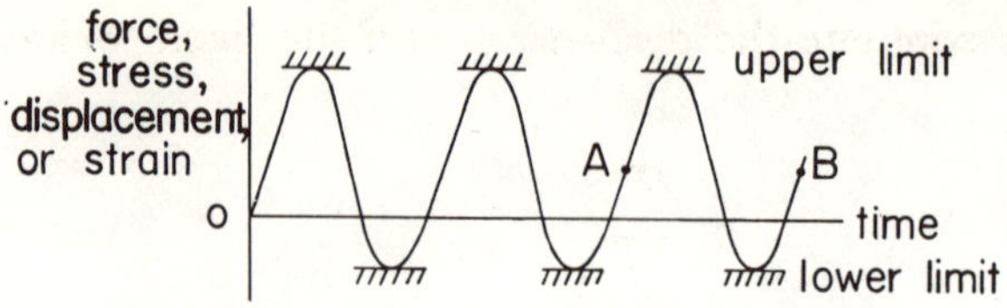

Fig. 22. Schematic of cyclic loading

to make the material meet the other wall, as in Fig. 22. A complete cycle such as that between points *A* and *B* thus includes two reversals in the direction of loading.

The following discussion will frequently refer to the notations of Figs. 23 and 24. Peak-to-peak values are the stress or strain

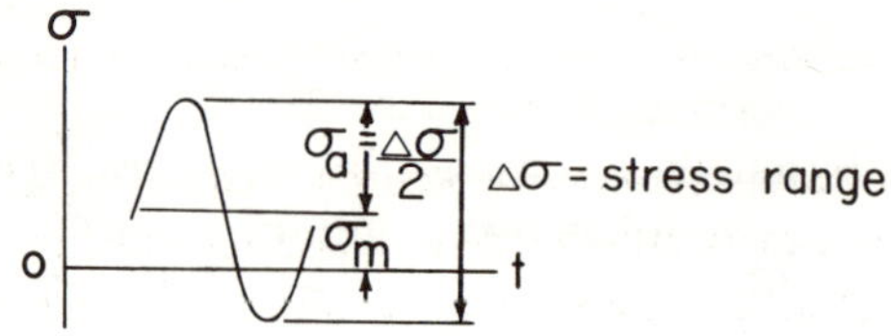

Fig. 23. Definition of cyclic stresses

ranges, and half-of-range values are the stress or strain amplitudes. The strain notation may be altered to indicate plastic strain, $\Delta\epsilon_p$, or elastic strain, $\Delta\epsilon_e$. The position of the halfway level between the peaks relative to the zero level is the mean stress, σ_m, or the mean strain, ϵ_m. Fully reversed conditions refer to either σ_m or ϵ_m, or both simultaneously, being zero. Unless otherwise specified, the fully reversed condition is used in this text.

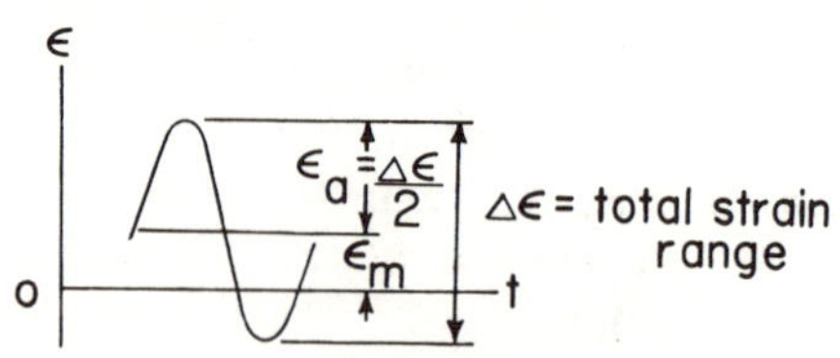

Fig. 24. Definition of cyclic strains

When a material is cyclically loaded within the elastic region, stress and strain are linearly related to each other by the elastic modulus, and it is necessary to measure only one of the two to know both of them (Fig. 25). An extremely serious limitation of this apparently simple case will be discussed later. For the time being let us continue the discussion with the linear behavior shown in Fig. 25.

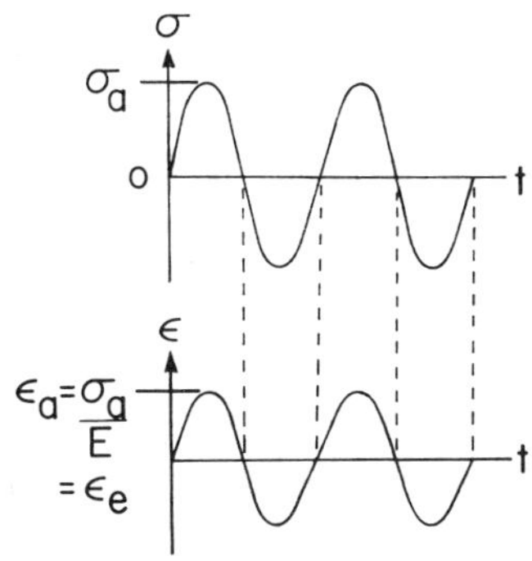

Fig. 25. Phase relation of cyclic stresses and strains for elastic behavior

It is possible to take the two functions from Fig. 25 and eliminate the time variable between them to produce strain as a function of cyclically varying stresses. This amounts to plotting cyclically changing stress vs. the strain it causes at every instant of time. Figure 26 shows how this can be accomplished using the linear part of a stress-strain curve. Assume that function 1 is a complete cycle of stress imposed on the material. Function 2 is the resultant cycle of strain whose values at each instant can be either calculated using the elastic modulus or obtained graphically by projections from the elastic line. For example, points B and D are horizontal components of projections of points A and C, respectively, on the elastic line. Naturally, one could reverse the process by imposing the cycle of strain shown and obtain the same cycle of stress as function 1.

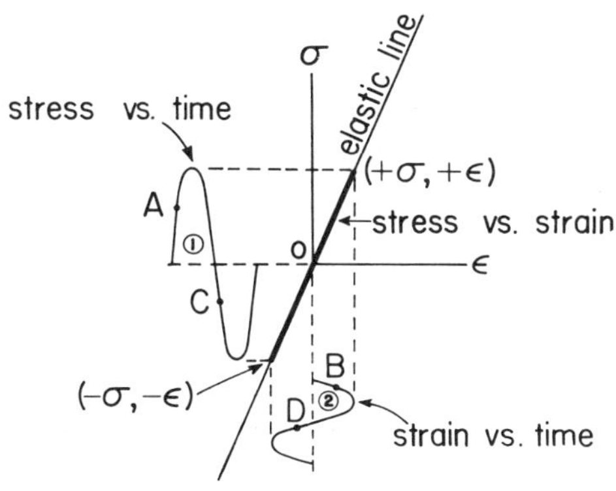

Fig. 26. Cyclic stress-strain plot for elastic behavior

An actual view of the process described above is provided by a simultaneous, interdependent recording of the stress and strain signals on an oscilloscope or an X-Y plotter. For either imposed function of Fig. 26 the stress vs. strain signal would vary from point 0 to $(+\sigma, +\epsilon)$ to 0 to $(-\sigma, -\epsilon)$ and back to 0. If the cycling were continued between the limits shown, the stress vs. strain signal would move back and forth along the heavy line in Fig. 26. We could say that points 0, $(+\sigma, +\epsilon)$, and $(-\sigma, -\epsilon)$ are on a cyclic stress-strain curve. This concept will be further developed.

Loads that produce inelastic deformation lead to more complex stress-strain responses than that of Fig. 26. The behavior is an extension of the quasi-static processes discussed earlier. Assume that the tensile and compressive stress-strain curve of a material is that shown in Fig. 27. Points A and B represent identical magnitudes of stress and strain. Figure 28 shows an idealized sequence

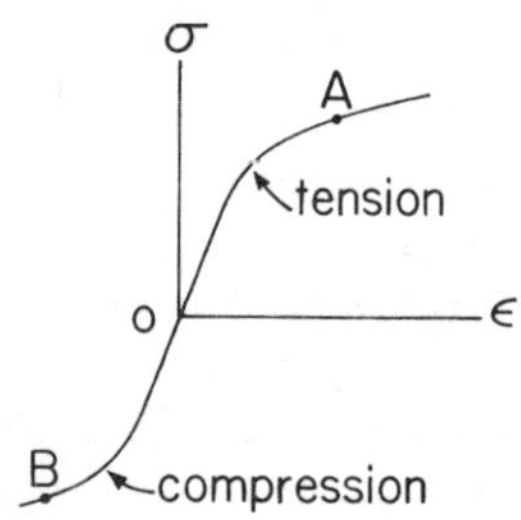

Fig. 27. Idealized tensile and compressive stress-strain curve

of loading for this material. From point 0 to A there is an increasing tensile load. We reach B by unloading from A and reversing the load into compression. Unloading from B and reversing the load into tension we reach point A again. Continuing the process we

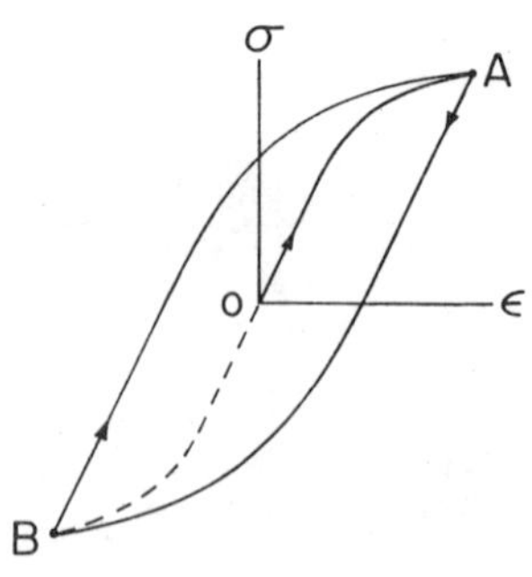

Fig. 28. Development of the hysteresis loop

repeatedly go from point A to B and back to A. One complete traverse along this route traces a curve (Fig. 29) called the hysteresis loop. On the basis of stress and strain, the hysteresis loop is the best way to describe the material's behavior during cyclic loading. It should be realized that regardless of the size of the diagram, the loop shows the events at a single point in the material. Thus the loops may be different for two infinitesimally small neighboring volumes of material.

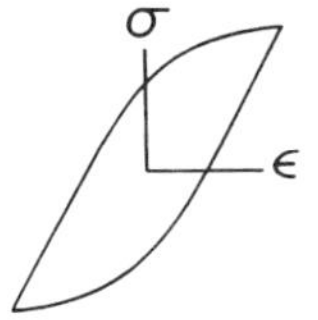

Fig. 29. A complete hysteresis loop

The most important property of a hysteresis loop is not its ability to show cyclically varying stresses but its ability to measure plastic strain per cycle. Cyclic plasticity is repeatedly reversed irrecoverable strain. This concept is explained by the aid of a model showing two rows of identical atoms (Fig. 30). Without external forces on them, the atoms are comfortable in their positions (part *a*). Under a shear force τ the parallel rows could slide relative to each other, if atom A can climb over the obstacle presented by B, and then by D, etc. Atoms of a row are comfortable in each trough between the atoms of the neighboring row. Assume the upper row "ratchets" to the right a distance of three atomic spacings (part *b*), where

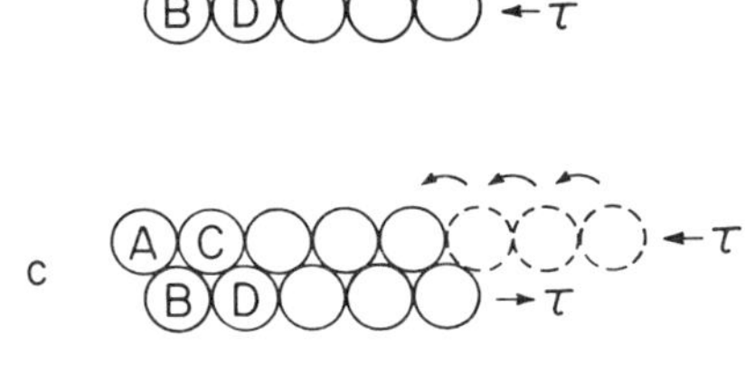

Fig. 30. Model of cyclically reversed plastic deformations

the shear force is removed. Since all of the atoms are comfortable in this position, the amount of displacement is irrecoverable, and we call it plastic strain. We can, however, apply a shear force in the opposite direction and push the upper row back to where it started from. Thus the final configuration (part *c*) is identical to that in (*a*), and the total plastic strain is six atomic spacings. Of course the end result is zero displacement and does not indicate the events that have preceded. For the complete cycle of loading in Fig. 30 the plastic deformation can be measured by the peak excursion of three spacings, or by the total distance of six spacings. The two values refer to the same event of a full cycle, but they can lead to some confusion and possible error, as will be seen later.

Once again, the model presented is highly idealized. In engineering materials the plastic deformation is strongly affected by the presence of dislocations. Cyclic deformations, in turn, may have a large effect on the number and distribution of dislocations. Such interactions are intimately involved in the mechanisms of fatigue damage initiation and accumulation. The details of these mechanisms are still being debated by researchers. It may suffice to say here that regardless of the exact mechanism, cyclic plastic strain is a measurable physical quantity that can be related to fatigue damage much better than can other factors present in the fatigue problem.

For a quantitative description of the hysteresis loop, consider the notation used in Fig. 31. Assume that the loop is symmetric with respect to the origin of the σ-ϵ coordinates. A complete cycle

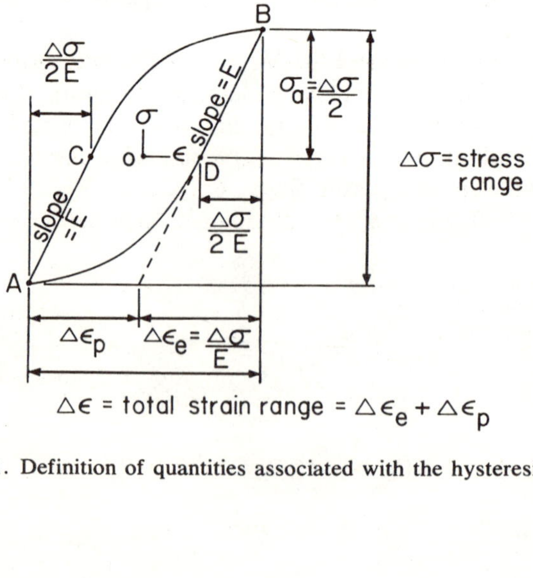

Fig. 31. Definition of quantities associated with the hysteresis loop

means starting from any point on the loop, tracing the loop in a clockwise direction, and ending up at the starting point. Regardless of the starting point or the control condition, the tips A and B of the loop represent the stress and strain limits of the cycle. The total strain is made up of elastic and plastic strain components as in static deformation. The total elastic strain per cycle is $\Delta\epsilon_e = \Delta\sigma/E$. Another way of saying this is that in each complete cycle there are two unloadings, and in each one the stress changes from a peak at point A or B to zero. Each recovered elastic strain is equal to $\Delta\sigma/2E$, according to Fig. 31, so that two of these are equal to $\Delta\epsilon_e$. The plastic strain range, $\Delta\epsilon_p$, is thus equal to $\Delta\epsilon - \Delta\epsilon_e$, and it is also shown rather vividly by the width of the loop at its midsection, between points C and D.

The above described notation follows generally accepted practice with respect to hysteresis loops, but it is not entirely correct. The difficulty comes from referring to various quantities as *per cycle*, whereas they could be obtained from loading through half a cycle. For example, it is not necessary to go around the loop completely to obtain $\Delta\sigma$ or $\Delta\epsilon$. Either one is given by going from A to B or from B to A. Consequently $\Delta\epsilon_p$ is not the plastic strain in a full cycle but in a half cycle. In other words, if we start from point A and stop at B, the accumulated plastic strain is $\Delta\epsilon_p$, and the same amount is experienced by the material in returning from B to A. This is analagous to the events described in Fig. 30. Nevertheless a completely drawn loop is very descriptive, and we can avoid other possible misunderstandings by using the diagram of Fig. 31 and referring to the important items as per-cycle quantities. The real meanings of these should be kept in mind for use in precise analytical work.

The area inside a hysteresis loop is the work done, or energy lost, per cycle. There should be no confusion about the calculation of this energy, since in this case the reversed plastic strains are automatically taken into account. One could integrate directly between the two separate paths from A to B in Fig. 31, or it could be done in two steps as shown in Fig. 32. The area between points

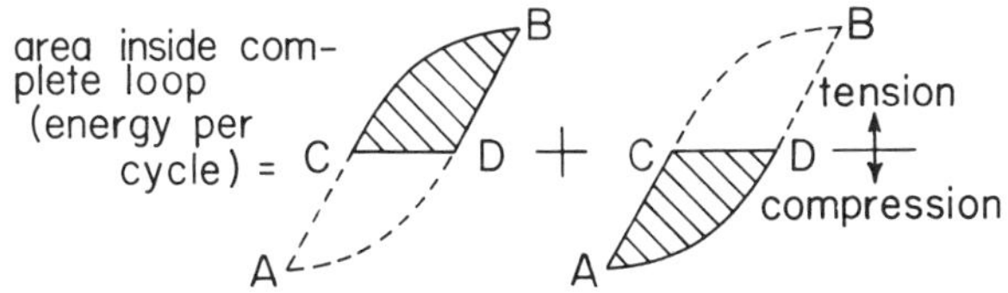

Fig. 32. The additive nature of hysteresis energy

C, B, and D is the work done by tensile stresses in creating plastic flow from C to D. The area between points D, A, and C is the work done by compressive stresses in plastically deforming the material from D back to C.

The monotonic stress-strain curve indicates roughly how much energy per cycle is dissipated for a given stress or strain range. At small stresses the loop is thin and the area is small, and at large stresses the loop is fat and the area is large. It appears that the entirely elastic behavior at small stresses, illustrated in Fig. 26, is a limiting case. Here the hysteresis loop has zero width and it degenerates into a straight line. As it is reasonable to expect some plastic flow even at near-zero stresses, however, we must assume a finite width of the hysteresis loop at all finite stresses. The heavy straight line in Fig. 26 is an extremely thin hysteresis loop whose two lines cannot be distinguished from each other, because of poor magnification of the strain function. The available instrumentation has this limitation, but it can be argued that materials fatigue is always caused by plastic strain that is at least locally present during every cycle. Because of the importance of cyclic plasticity in fatigue, relatively fat hysteresis loops will frequently be used for purposes of discussion even when their sizes are not realistic.

The practical aspects of cyclic stress and strain measurement or control range from the near impossible to those that are manageable. For example, the measurement and control of cyclic strains in the neighborhood of complex geometries such as notch roots and other sudden changes in cross-sectional area is not yet possible. Relatively simple specimens can be used to produce meaningful results, but only because in these the stresses and strains do not change abruptly. Figure 33 shows a typical simple laboratory specimen intended for axial push-pull testing. The enlarged sections are for gripping. The uniform reduced section is for stress-strain measurements and for the localization of the eventual failure. The assumptions made in analyzing the behavior of this specimen are that (1) every cross section in the reduced part of the specimen has the same uniform stress on it, and (2) the axial strain is the same at every point in the reduced section. When these assumptions are valid it is possible to record true hysteresis loops for the material, even though a loop applies only to an infinitesimal volume of material —something that we cannot yet record. What can be done is a measurement of average stress at the minimum cross-sectional area and a measurement of strains averaged over a sizeable length some-

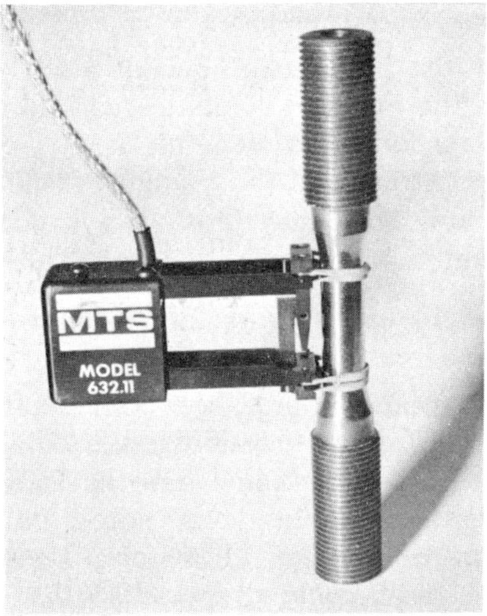

Fig. 33. Laboratory specimen with extensometer for longitudinal strain measurement

where in the reduced section. Diametral strains can also be used to plot hysteresis loops for very short or otherwise limited specimens. Figures 33 and 34 show examples of strain-measuring extensometers attached to laboratory specimens. For further details about modern testing procedures see [42].

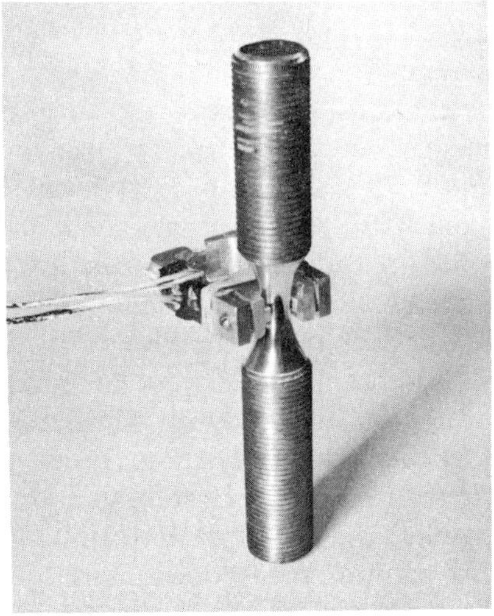

Fig. 34. Laboratory specimen with diametral clip-gage extensometer

2.2 Cycle-Dependent Material Response

Discussions of the hysteresis concept already indicate a certain manner in which the material wants to respond to a specific loading condition. This response, in general, is not constant from cycle to cycle. The two extremes of the changing response are cycle-dependent hardening and softening (not always corresponding directly to standard indentation hardness values). These will be discussed with respect to the imposed conditions.

Stress Control

The possible responses range between the two extremes shown in Fig. 35. The controlled function has a constant amplitude, σ_a. The independent variable, strain, has the same frequency as the imposed load, but its amplitude is not a constant. In the case of cycle-dependent hardening, the material's resistance to deformation is increasing from cycle to cycle so that the strain becomes smaller and smaller under the same stress. It will be shown later that the tangent modulus increases in this case. The process is similar to cold-working in monotonic deformation. The envelope of the strain peaks (dotted line in Fig. 35*b*) is an exponential function with most of the change taking place early in the cycling. Normally, very little further adjustment occurs past half the fatigue life at a given stress.

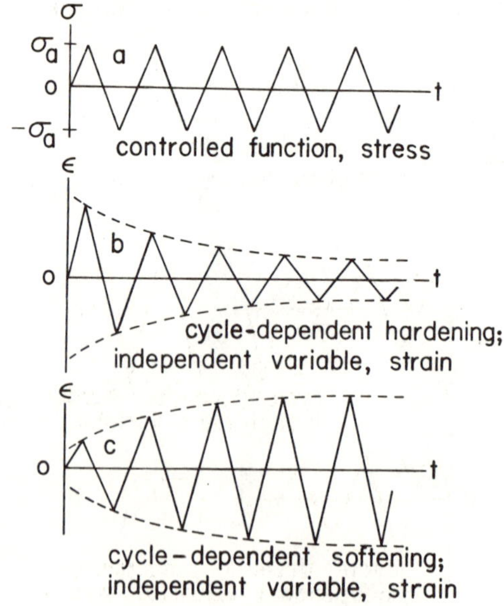

Fig. 35. Cycle-dependent material responses under stress control

Cycle-dependent softening is a gradually decreasing resistance to deformation. Under stress control this is evidenced by strains increasing from cycle to cycle under an exponential evelope. The situation here can easily become unstable if the strain per cycle becomes a large fraction of the strain at fracture in monotonic tension. The problem may be compounded by the controlled function itself, since true stress control is easy only at small deformations, and unmanageable when cracks form in the material. Thus in some cases (necking, presence of cracks, imperfections) the control function is essentially load control, even when stress control is intended. An analogy of this problem is a dead weight suspended by a wire. The weight is held up by the force developed in the wire, not by the level of the stresses in it which depend on the possibly changing size of the cross-sectional area. In cycling, too, stresses are known only as well as the force and the area are known.

Strain Control

Again the controlled function has a constant amplitude, ϵ_a in Fig. 36*a*. The independent variable, stress, changes under exponential envelopes in both extremes shown. Cycle-dependent hardening (Fig. 36*b*) means that the resistance to deformation increases so that larger and larger stresses are required to push and pull the material to the unchanging strain limits. This is not a runaway condition in the sense that softening under stress control is, in spite of the

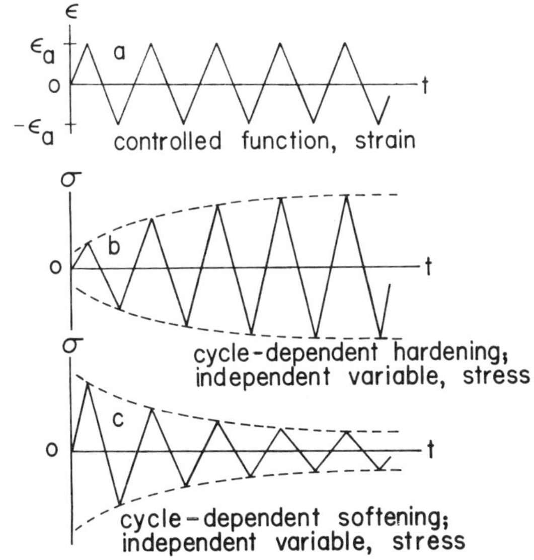

Fig. 36. Cycle-dependent material responses under strain control

similar envelopes. Problems in this case can arise if the material (high strength, brittle) is sensitive to high levels of stress.

Cycle-dependent softening makes the material deform more easily. Under strain control, therefore, the stress needed to enforce the strain limits becomes smaller and smaller as cycling progresses (Fig. 36*c*). As in stress control, very little further adjustment in the independent amplitudes occurs past half the life at a given cyclic strain level.

2.3 Hysteresis Loops with Cycle-Dependent Response

The situations discussed above can also be represented on stress vs. strain diagrams. In this section extensive use is made of the idea introduced earlier that the control limits are walls between which the material is forced back and forth. At the same time, the material is allowed to respond on its own by "hitting" the walls wherever it wants to.

Under stress control the walls are horizontal, as in Fig. 37. The

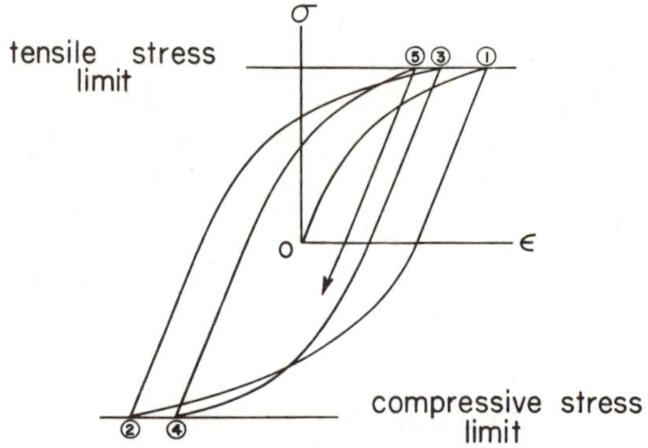

Fig. 37. Stress vs. strain for stress control and cycle-dependent hardening

material may be loaded in the tensile direction starting from the origin of the stress-strain coordinates. The tensile limit is reached at a strain denoted by the number 1. From 0 to 1 the curve coincides with the monotonic stress-strain curve. After 1 the material experiences a reversal in the loading direction which sends it toward point 2, and so on. The successive numbers indicate the strains at which the load limits are reached in the sequence. It is somewhat imprecise to talk about hysteresis loops when the loop never returns exactly to where it started from, but normally the differences are small enough to allow that. Also, an imperfect hysteresis loop is still good enough to show the behavior of the material. The important thing is that the width of the loops ($\Delta\epsilon_p$) is gradually decreasing with successive cycles until a fairly stable loop shape develops somewhere before the half-life.

Cycle-dependent softening under the same type of stress control is shown in Fig. 38. Again, the control limits are reached at strains

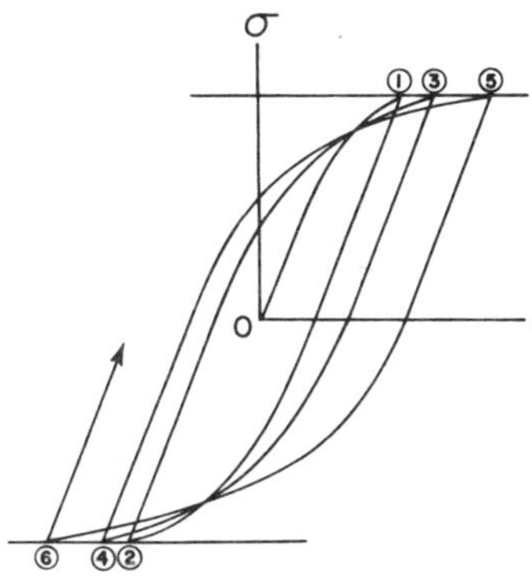

Fig. 38. Stress vs. strain for stress control and cycle-dependent softening

denoted by the numerals, but now the strains increase from cycle to cycle. The runaway nature of the process is clearly indicated by Fig. 39, which was reproduced from test data. It is striking that the stress amplitude is smaller than the tensile yield strength.

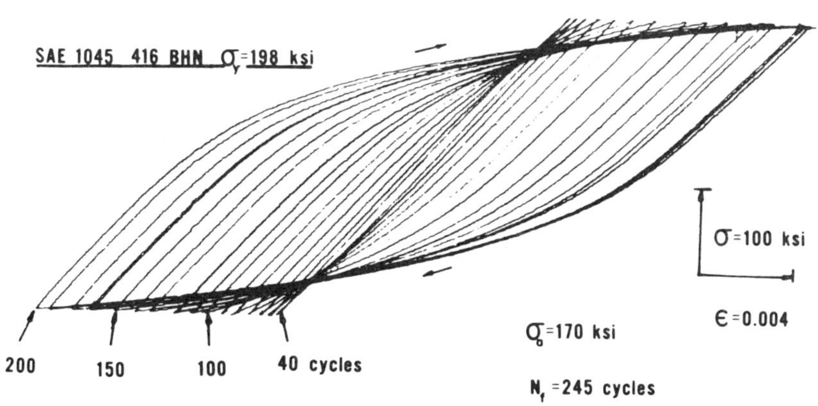

Fig. 39. Example of hysteresis loops for stress control and cyclic softening (from Morrow, Halford, and Millan, "Optimum Hardness for Maximum Fatigue Strength of Steel")

Vertical walls are used to show strain control, where the points (stress levels) at which the walls are reached depend upon the material's response. Cycle-dependent hardening is described in Fig. 40. The successive numbers represent the stress levels at which

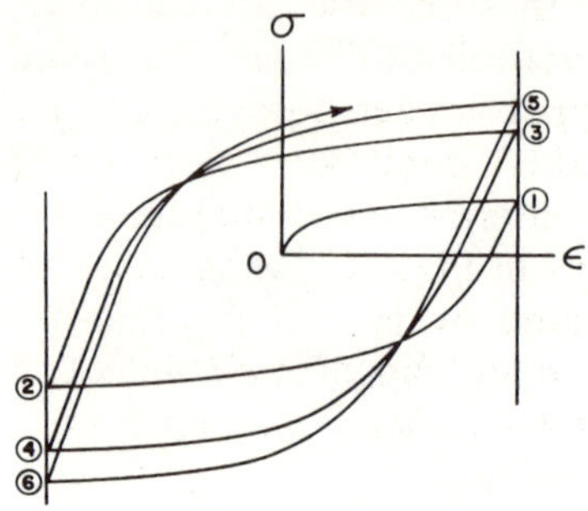

Fig. 40. Stress vs. strain for strain control and cycle-dependent hardening

the strain limits are reached as cycling progresses. Since hardening means an increasing resistance to deformation, the stresses required to push and pull the material from one strain limit to the other gradually increase. The unloading from each loop tip is along the elastic slope so that the plastic strain per cycle is decreasing with successive cycles.

Cycle-dependent softening under strain control is shown in Fig. 41. Here the material's resistance to deformation gradually decreases and the stresses required to enforce the strain limits become smaller and smaller. At the same time the plastic strain per cycle increases.

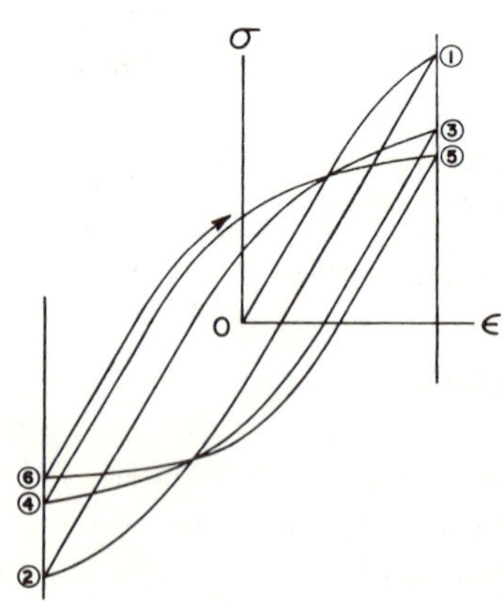

Fig. 41. Stress vs. strain for strain control and cycle-dependent softening

2.4 The Qualitative Prediction of Material Response

Whether a certain material cyclically hardens or softens depends mainly on the initial condition of the material, and in some cases also on the magnitudes of the imposed cyclic loads. Concerning the initial condition, it is observed that a given material may either harden or soften cyclically. This is illustrated in Fig. 42, where annealed OFHC copper is seen to cyclically harden at a rapid rate, but in the cold-worked condition to cyclically soften. The same metal could have an intermediate initial hardness allowing only small cyclic adjustments, as in Fig. 42*b*. In the case of metals, the intermediate range is indicated by an initial strain-hardening exponent, n, being in the approximate range of 0.10 to 0.20. When n is above this range, the metal cyclically hardens, and n below this range indicates cyclic softening.

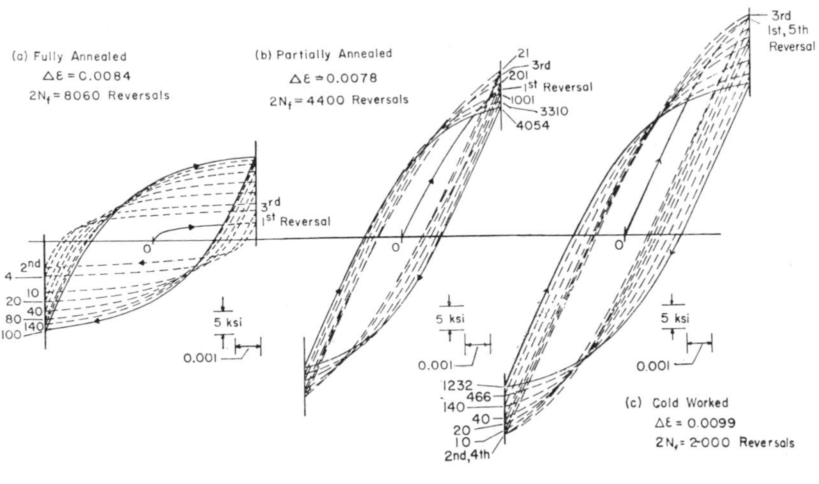

Fig. 42. The role of initial material condition in cyclic stress-strain response: (a) fully annealed, showing cyclic hardening; (b) partially annealed; (c) severely cold-worked, showing cyclic softening (after Tuler, in Morrow, *Internal Friction, Damping, and Cyclic Plasticity*. Courtesy ASTM)

The different responses can occasionally be seen in a given metal as a result of different cyclic loads. For example, consider mild steel whose monotonic stress-strain curve is shown in Fig. 43. Under cyclic stress amplitudes (σ_a) that are smaller than the tensile yield (σ_{ty}) this steel has a tendency to cyclically soften. Cyclic stresses that are within the cold-working range cause hardening in the same steel.

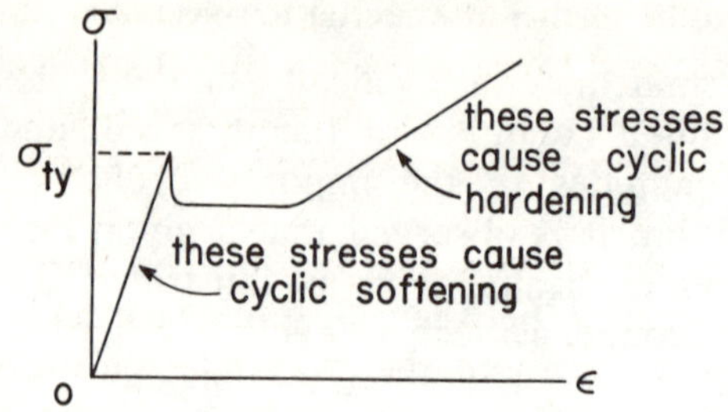

Fig. 43. Cyclic response of mild steel with respect to the load range

2.5 Cycle-Dependent Changes in Mechanical Properties

On the basis of the foregoing it is reasonable to consider carefully the possible changes in mechanical properties that cyclic loading may cause [43, 64]. Another convincing argument for looking for such changes is provided by the following relatively simple experiments. Take several metal specimens of the same composition, size, shape, and initial condition. Some of these can be used to establish the monotonic stress-strain curve as the solid line in Fig. 44. Other specimens from the same group receive some cyclic loading, perhaps to half the fatigue life at a given amplitude. After the cycling, the specimens are returned to the state of zero stress and zero strain (it will be shown later how this is done). Now the only difference between these specimens and the virgin ones is that these have cyclic deformations in their histories. The specimens thus treated are tested in monotonic tension, and the stress-strain curves are plotted together with the virgin specimens' data. It is possible that the curves for the two groups of specimens will coincide, which means that the metal is cyclically stable. A curve higher

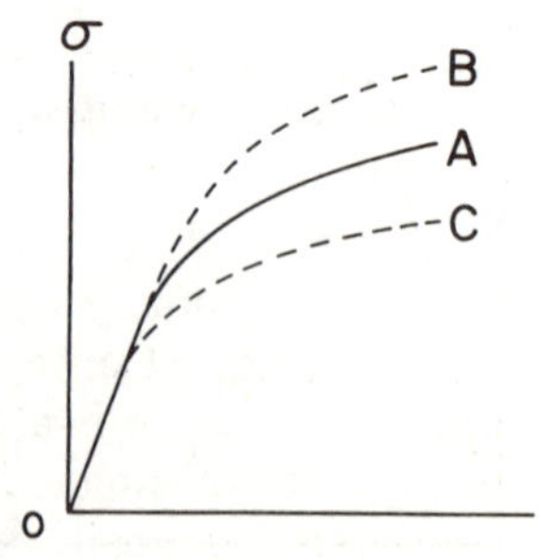

Fig. 44. Cycle-dependent changes in stress-strain response. A is monotonic curve, B, cyclic hardening, and C, cyclic softening

than in the original monotonic tension indicates that the material cyclically hardens; one lower indicates that it cyclically softens.

The stress-strain curves found in the above-described procedure are appropriately referred to as the cyclic stress-strain curves. There are a number of methods [34, 37] that can be used for the determination of the cyclic stress-strain curve, and there are small differences in the results between different methods. The essential concern in all cases is about the material's deformation resistance (or tangent modulus) after it has endured cyclic loading but has not yet failed. Changes in the deformation resistance can be enormous, and of course the practical significance of such changes in fatigue design is substantial. Figures 45 and 46 show the monotonic and cyclic stress-strain behaviors of several metals. To appreciate the message of these plots, let us consider in detail the last diagram (380 BHN) in Fig. 46. The yield strength in monotonic tension is about 200 ksi, and it would appear safe to specify in design an operating stress of 150 ksi. Not too long ago all design procedures were based on this kind of a concept of safety from failures, and even today much designing relies on such ideas. For the steel used as an example, however, the designer could be terribly wrong.

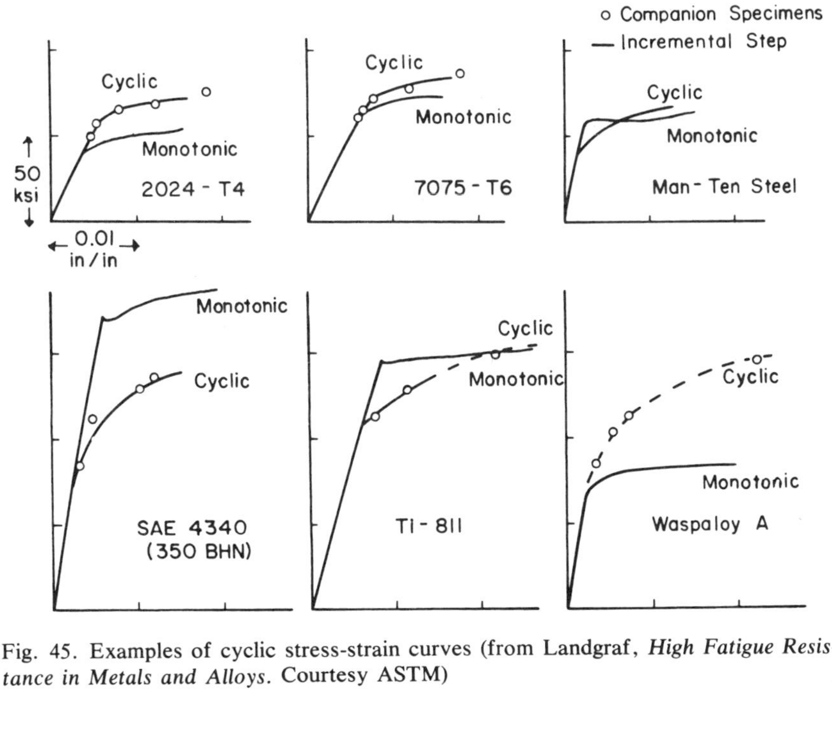

Fig. 45. Examples of cyclic stress-strain curves (from Landgraf, *High Fatigue Resistance in Metals and Alloys.* Courtesy ASTM)

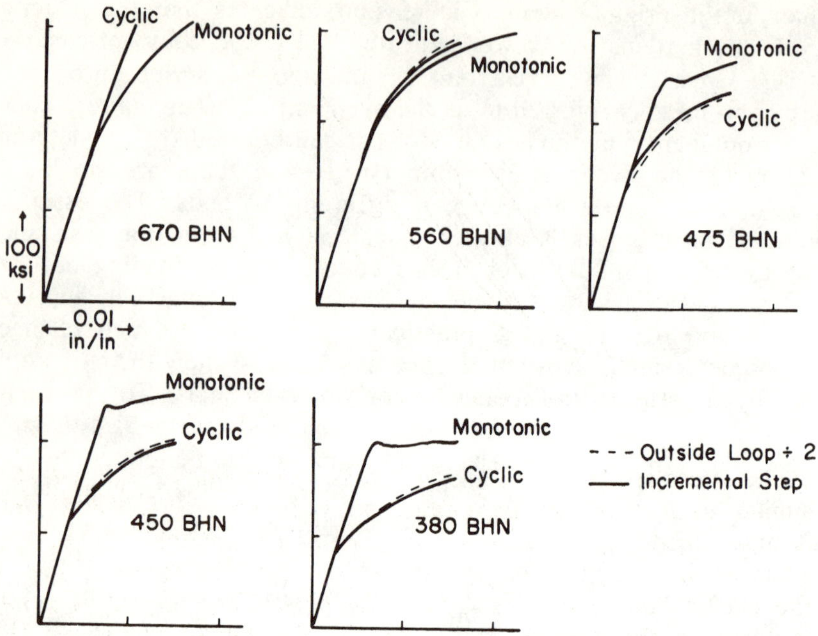

Fig. 46. Influence of hardness on the cyclic stress-strain response of SAE 4142 steel (from Landgraf, *High Fatigue Resistance in Metals and Alloys*. Courtesy ASTM)

He would expect about 0.005 in/in of completely elastic strain at a stress of 150 ksi, but the material might be responding along the cyclic curve. There the strain would be twice his expected value, and half of this would be plastic strain. Besides the possible problem of having too much strain, the situation could gradually worsen under cyclic load control (runaway condition for cyclic softening) and result in premature failure. The conclusion that must be remembered from the above is that *in general, the strain produced by a given stress depends (among other things) on the prior cyclic-loading history of the material.*

A rapid method of determining the cyclic stress-strain curve is by the multiple-step test [34, 37]. The test can be started at a small strain amplitude, for example, and the amplitude increased in small increments. At each level of strain enough cycles are performed to produce a stable hysteresis loop (but not too many to cause serious fatigue damage), and the loop is recorded. All of the successive loops are plotted on the same diagram with a common σ-ϵ coordinate system. A smooth curve is drawn to connect the tips of these stable loops, as shown in Fig. 47. This curve is the

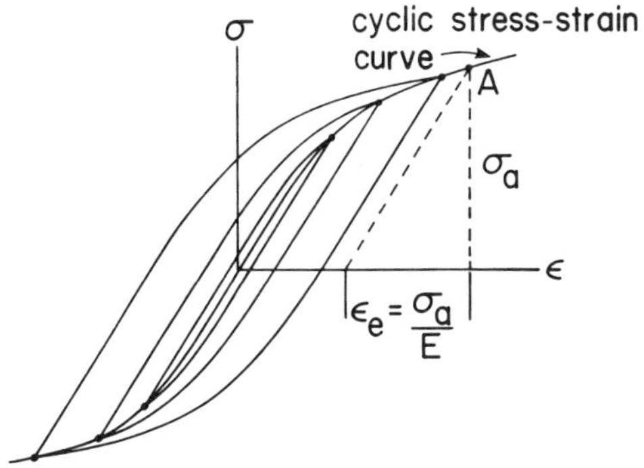

Fig. 47. One definition of the cyclic stress-strain curve

cyclic stress-strain curve, and it can be represented by an equation similar to that of the monotonic stress-strain curve. Total strain is again made up of elastic and plastic components: $\epsilon = \epsilon_e + \epsilon_p$. Thus at any point A on the cyclic stress-strain curve in Fig. 47 the elastic component is σ_a/E. The plastic component is similar in form to the plastic strain part of the monotonic strain (Eq. 11). For the static case,

$$\epsilon_p = \left(\frac{\sigma}{K}\right)^{1/n} = \epsilon_f\left(\frac{\sigma}{\sigma_f}\right)^{1/n} \tag{13}$$

since $K = \sigma_f/\epsilon_f^n$, meaning that $\epsilon_p = (\sigma/K)^{1/n}$ is also valid at the instant of fracture, where $\sigma = \sigma_f$ and $\epsilon_p = \epsilon_f$. We can expect that the plastic part of the cyclic stress-strain curve can be expressed in a form similar to Eq. 13, but that the constants K, ϵ_f, σ_f, and n might be different. To reflect the similarity of concepts between the static and cyclic stress-strain properties, the constants referring to the cyclic behavior are denoted ϵ_f', σ_f', and n'. They are called the fatigue-ductility coefficient (ϵ_f'), the fatigue-strength coefficient (σ_f'), and the cyclic strain-hardening exponent (n'). If information to the contrary is not available, the following assumptions can be made: $\epsilon_f' \approx \epsilon_f$, and $\sigma_f' \approx \sigma_f$. (For summaries of various methods of estimating ϵ_f' and σ_f' see [34].) Changes from n to n' are such that (for metals) an initially low n ($n < 0.10$) will increase, and an initially high n ($n > 0.20$) will decrease. The method of finding values of n' is shown in Fig. 48. The effects of cycling on important mechanical properties are summarized in Table 1.

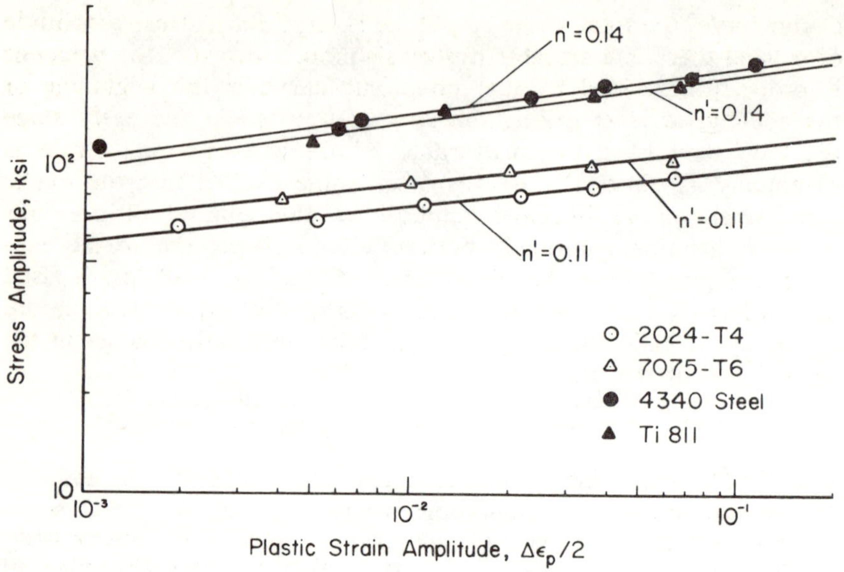

Fig. 48. Experimental determination of the cyclic strain-hardening exponent (from Endo and Morrow, "Cyclic Stress-Strain and Fatigue Behavior of Representative Aircraft Metals." Courtesy ASTM)

The final general mathematical expression relating cyclic strains and cyclic stresses is in the form of $\epsilon = \epsilon_e + \epsilon_p$ [48]:

$$\frac{\Delta\epsilon}{2} = \frac{\sigma_a}{E} + \epsilon_f'\left(\frac{\sigma_a}{\sigma_f'}\right)^{1/n'} \tag{14}$$

The total strain amplitude equals the sum of the elastic strain amplitude and the plastic strain amplitude. It must be remembered that Eq. 14 is not entirely precise, because it describes the behavior of a material in the cyclically stabilized state. By definition there is rarely if ever (possibly for initially stable materials) such a thing as a stabilized state. To illustrate this, consider the monotonic and cyclic curves for a softening metal in Fig. 49. According to the

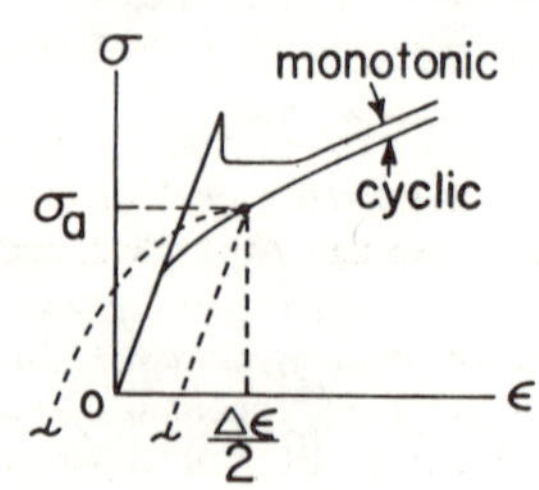

Fig. 49. Gradual change from monotonic to cyclic stress-strain curve

cyclic curve, the total strain amplitude is $\Delta\epsilon/2$ for a stress amplitude of σ_a and there is a sizeable hysteresis loop. However, the behavior is properly described by the monotonic curve in the beginning of the cycling, at least during the first half-cycle. At the early stage the loop may have no measurable width, and the total strain is essentially all elastic. It is only after some cycling that the cyclic stress-strain curve becomes applicable. The number of previous cycles is arbitrary, normally between 20 to 50 percent of the estimated fatigue life at the given level of loading. Half-life is used most often for convenience, and because the exponential nature of the cyclic adjustments generally allows only little change in the second half of the life.

Table 1. Cycle-dependent changes in the mechanical properties

Mechanical property	Effect of cycling
E, modulus of elasticity	Negligible effect
Hardness S_{ty}, tensile yield point[a] S_{ys}, offset yield strength S_u, ultimate strength	These are flow properties, since they are related to the ease with which plastic deformation can occur, and are all potentially changed by cycling. They will increase if the metal cyclically hardens, decrease if it cyclically softens, or remain unchanged if the metal is initially in a stable state. The initial n and the ratio S_u/S_{ys} can be used to estimate whether a metal will cyclically soften or harden.
σ_f, true fracture strength ϵ_f, true fracture ductility $\% RA$, percent reduction in area	These properties may be classed as fracture properties, since they are measures of the fracture resistance. They may be changed somewhat by cycling but not to the degree that the flow properties are changed.[b]
n, strain hardening exponent	This is also a flow property and can be greatly changed by cycling. All metals seem to cyclically adjust to a cyclic strain hardening exponent of $0.10 < n' < 0.20$. If n is initially low (a severely hardened metal) n will increase, if n is high it will decrease. Within the range 0.10 to 0.20, n will change only slightly.

Source: Raske and Morrow, in *Manual on Low Cycle Fatigue Testing.* Courtesy ASTM

[a] The yield point phenomenon may be removed by cyclic stressing. After a number of cycles below the yield point the hysteresis loop may open and the yield point disappear, indicating a cyclic softening. At higher cyclic strains the same metal may harden.

[b] There have been a number of investigations of "exhaustion of ductility" and loss of "residual strength" of fatigued specimens and components. It has been found that the strength and ductility are not drastically reduced until the later stages of fatigue where relatively large cracks are present.

The lesson from all this is that for a given material, the uniqueness of the cyclic stress-strain curve depends on the arbitrary definition of stabilization in stress-strain response. In reality there are numerous cyclic stress-strain curves, as shown in Fig. 50. At the present

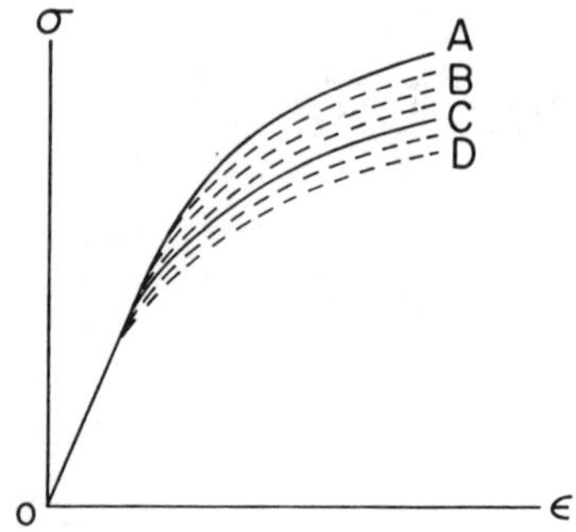

Fig. 50. Uncertainty of stabilization in stress-strain response: *A* is monotonic curve, *B*, response before half-life, *C*, "stable" response at half-life, and *D*, response after half-life

time, or in the foreseeable future, the available data will not allow such detailed descriptions of cyclic stress-strain response. In most practical situations this may not be necessary. What is needed, however, is a general appreciation of the strong chance that cycling can alter the stress-strain response. For this purpose the curve obtained at half-life stabilization, or even less than that, is an extremely effective aid.

3

The Fatigue Life

Fatigue-life data has been accumulated in large quantities since the second half of the last century. Most of the testing produced intermediate- and long-life data plotted as stress amplitude vs. the resulting life, commonly referred to as *S-N* curves. Figure 51 shows such a plot schematically. In steels, the life steadily increases with decreasing stress until the stress level of the fatigue limit is reached,

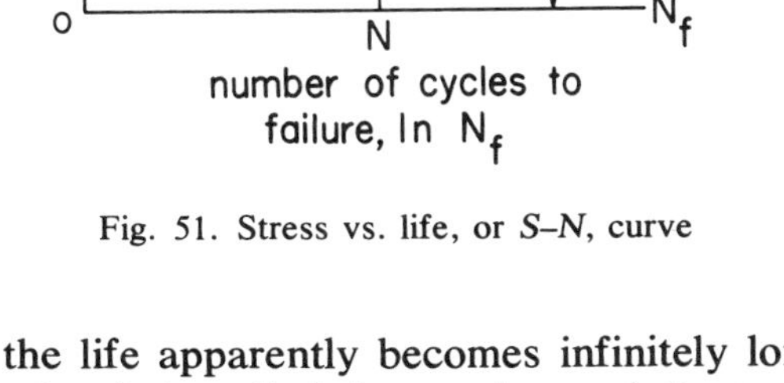

Fig. 51. Stress vs. life, or *S–N*, curve

below which the life apparently becomes infinitely long. In nonferrous metals such a fatigue limit is not observed. Instead, the concept of the fatigue strength is used: the stress level at which the metal will "live" a specified number of cycles (normally 10^8). Fatigue strength is an arbitrary description of a material's ability to withstand cyclic loading. For example, one could say that the fatigue strength at 100 cycles is 80 ksi, and at 10^7 cycles is 12 ksi, both for the same metal.

The concept of a real fatigue limit is somewhat dubious. The fact that not all metals show such a limit may indicate that possibly it is a fictitious quantity and that some metals simply have not been or cannot conveniently be tested to large enough cycles. In ferrous metals at small stresses the strain-aging mechanism tends to delay or offset the damaging effect of cyclic plastic strains. Whether the result is infinite life or not is immaterial. The important point is that in place of the earlier preoccupation with the fatigue limit, there should be more time spent in understanding the stress-strain concepts that have unifying significance in all of fatigue.

In dealing with the *S-N* curve, at least a brief mention must be made of scatter in the individual data points that make up the curve. Suffice it to say here that the scatter increases with increasing life range [63]. It is quite common to observe factors of two or three differences in life under the same stress level in the 10^6 cycles range. The reason this is discussed here is to bring attention to the exponential nature of the fatigue problem. A life of, for instance, two million cycles instead of an expected one million sounds grossly inaccurate, but on a logarithmic plot, as in Fig. 51, such a deviation does not appear to be serious. Life plots are normally on a logarithmic basis, and the scatter has to be studied in relation to the overall pattern of the life diagram. Other important aspects of the fatigue behavior that show exponential character are the rates of cyclic hardening and softening discussed previously, and the acceleration of crack propagation shown in Fig. 52.

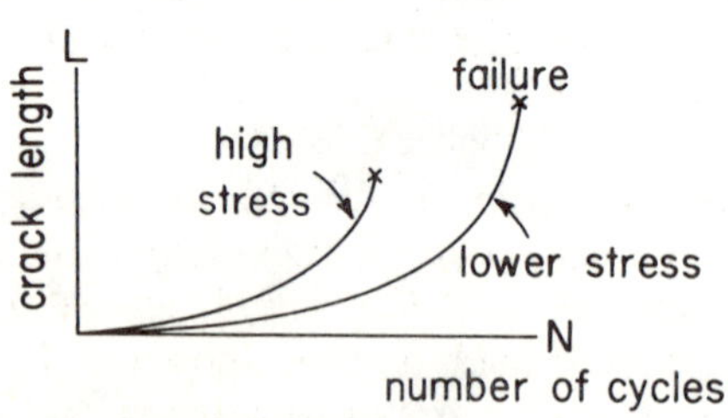

Fig. 52. Rates of fatigue-crack growth

In the 1950s, the field of low-cycle fatigue-testing (<50,000 cycles) was established and has grown steadily since [3, 5, 40, 42]. For dealing with problems within the short-life category, the *S-N* curve is not adequate. Controlled strain is found to approximate best the conditions existing in practice, and to avoid instabilities and runaway situations common under high levels of cyclic stress control. A typical curve of total strain vs. life is shown in Fig. 53. The qualitative statements made in this diagram concerning the degree of plasticity as a function of life range are not at all different from what one would expect on the basis of the monotonic behavior or the *S-N* curve. These qualitative aspects are discussed here,

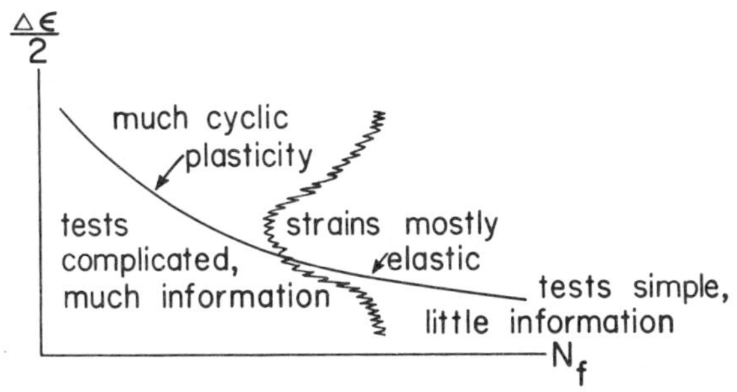

Fig. 53. Strain vs. life curve

before a quantitative analysis is attempted, because they directly affect our understanding of the fatigue problem. Many individuals who investigate and work on fatigue believe that there is little reason to understand low-cycle fatigue as long as their main interest is in high-cycle fatigue. There are a number of arguments for separate mechanisms of damage at the two life ranges, but from the mechanics view of stress and strain a unified picture emerges. That is, *fatigue damage is caused by cyclic plastic strain, and the only significant difference between various life ranges is the degree of cyclic plasticity.* Some evidence of this has already been mentioned, and more will be said about it later.

It must be added that some reluctance in learning about low-cycle fatigue is caused by the relative complexity of the subject. Data acquisition, for example, requires sophisticated testing, measuring, and recording equipment such as that shown in Fig. 54. In comparison with rotating or plate-bending machines and associated procedures, the present-day investments of financial and human resources to study fatigue do indeed appear formidable. The returns in knowledge, however, amply justify these efforts, and both the low- and high-cycle areas can benefit from the broad understanding that becomes possible.

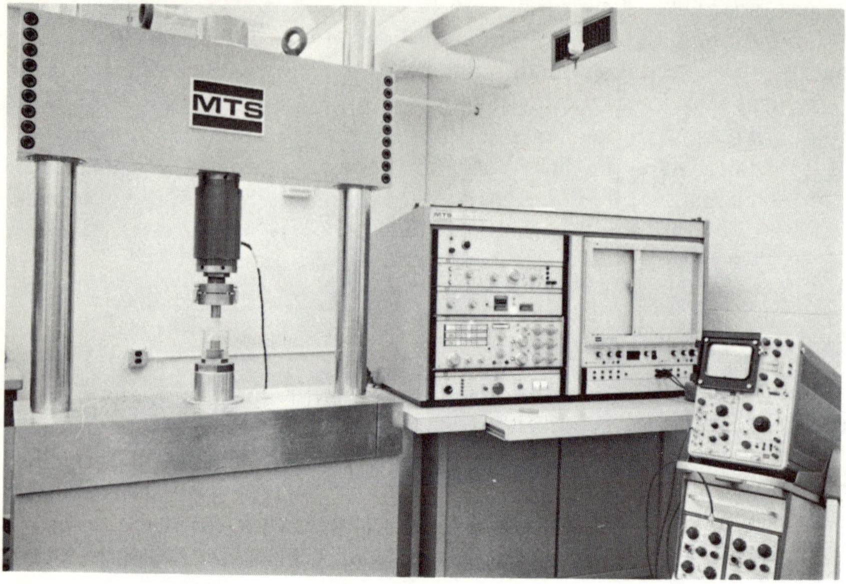

Fig. 54. Test equipment for the control and measurement of cyclic stresses, strains, or deflections

3.1 Analysis of the Strain-Life Curve

The fatigue life in number of cycles is normally given as N_f but in some cases it is useful to think of cycles as resulting from reversals in load, so that there are always two reversals to count in a complete cycle. The life in units of reversals is $2N_f$, an abbreviation which will be frequently used here. Of course the slope of a curve on the logarithmic plot is not affected by the factor of two conversion.

The analysis of a total strain vs. life curve (Fig. 53) is based on Eq. 14. Every value of total strain is made up of elastic and plastic components, so that at any given life these two components are present. An example of this is given with the aid of Fig. 55. Assume that the fatigue life is 10,000 cycles under a total strain amplitude of 0.6 percent. From an actual measurement it may be known that the plastic strain amplitude is one-third of the total, which leaves two-thirds, or 0.4 percent, for the elastic strain amplitude. We could then argue the other way around and claim that, for the same material, whenever the cyclic elastic strain amplitude equals 0.4 percent the life will be 10,000 cycles. Also, knowing only that the cyclic plastic strain amplitude is 0.2 percent, we could claim that the life is 10,000 cycles.

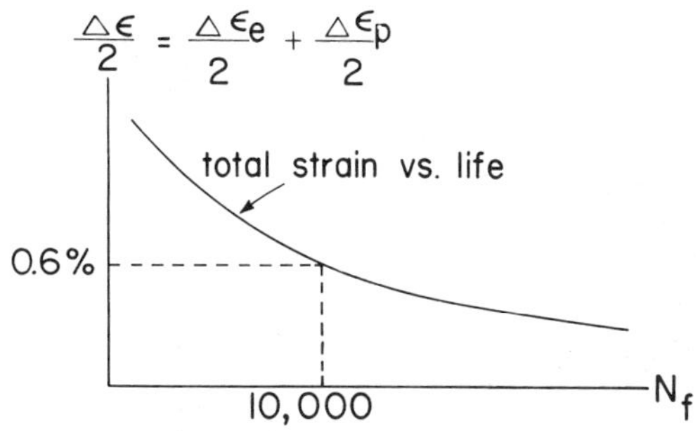

Fig. 55. Example of the cyclic strain components

Thus, for the range of all lives, each particular life number can be reached by specifying either a unique elastic or a plastic strain level. Consequently, curves may be drawn to show elastic strain vs. life and plastic strain vs. life on the same plot with total strain vs. life. On logarithmic plots the elastic and plastic curves appear to be fairly straight lines, as in Fig. 56. These curves are not

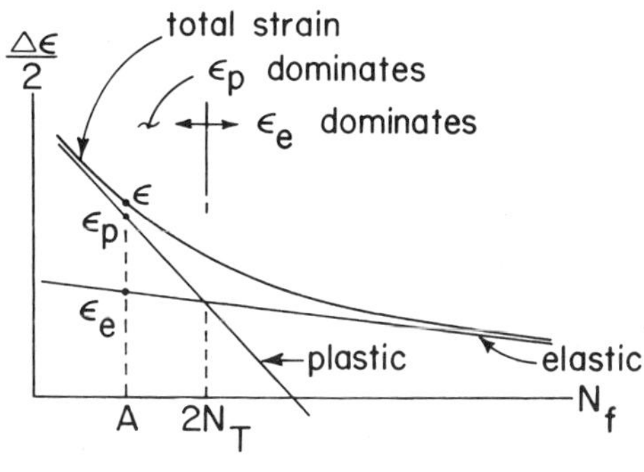

Fig. 56. Superposition of the strain components

independent of each other, since Eq. 14 applies. For example, a vertical line drawn at the life *A* relates the particular strains. Each strain amplitude, ϵ_e, ϵ_p, and ϵ, is with respect to zero strain, which is not shown on the logarithmic coordinate axis. Therefore the equation $\epsilon = \epsilon_e + \epsilon_p$ is not obvious in the graphical representation of Fig. 56, but the correct superposition of ϵ_e and ϵ_p on each other produces the total value ϵ.

There are a number of important properties that can be obtained from a compound diagram like that in Fig. 56. We find the so-called fatigue ductility properties in the plastic strain component vs. life curve (Fig. 57*a*), and the fatigue strength properties from the elastic strain component vs. life plot. The latter plot is readily converted to give stress vs. life for the direct representation of strength parameters, as in Fig. 57*b*. At one reversal we are dealing with a monotonic tension test, not strictly with fatigue, so the fracture properties shown at this "life" are ϵ_f and σ_f. Yet if there is a close relationship between fatigue and monotonic fracture, as one always hopes, it should be possible to extrapolate the cyclic curves and find that the respective intercepts ϵ_f' and σ_f' coincide with ϵ_f and σ_f. For the example in Fig. 57 the agreements are very good, but we are not always that fortunate. Figures 58 and 59

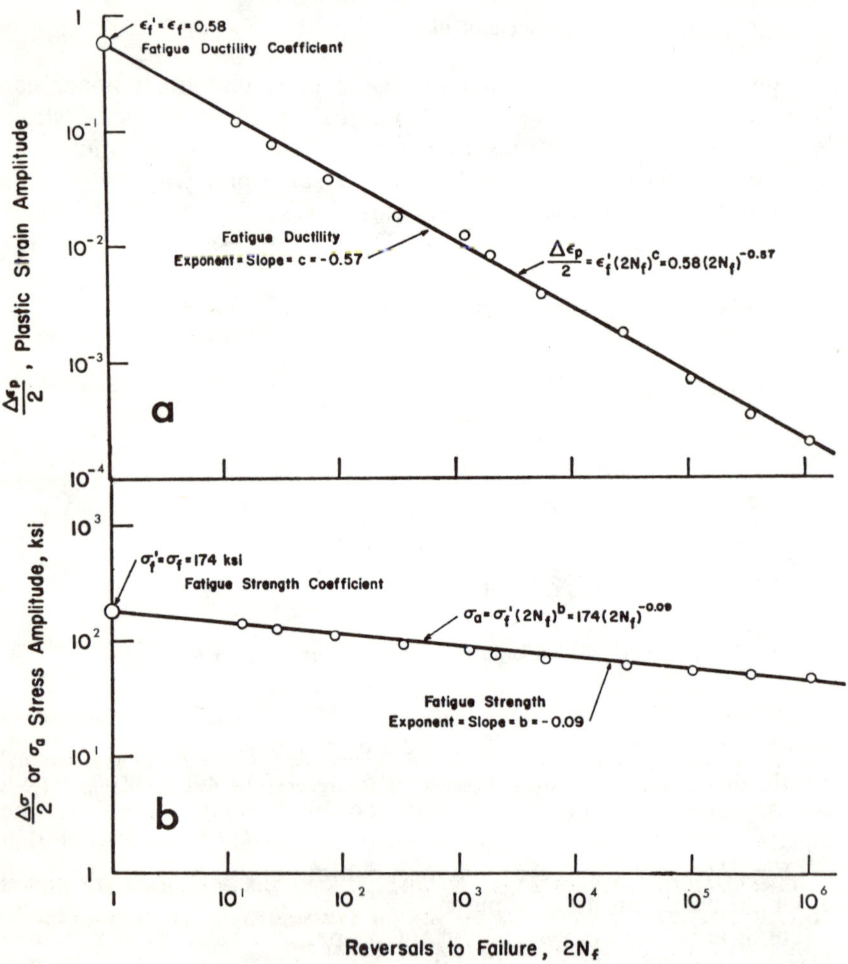

Fig. 57. Fatigue properties of SAE 4340 steel: (a) fatigue-ductility properties; (b) fatigue-strength properties (from Morrow, *Internal Friction, Damping, and Cyclic Plasticity*. Courtesy ASTM)

show that the linear extrapolations to one reversal (giving values of ϵ_f' and σ_f') do not necessarily agree with the monotonic data ϵ_f and σ_f.

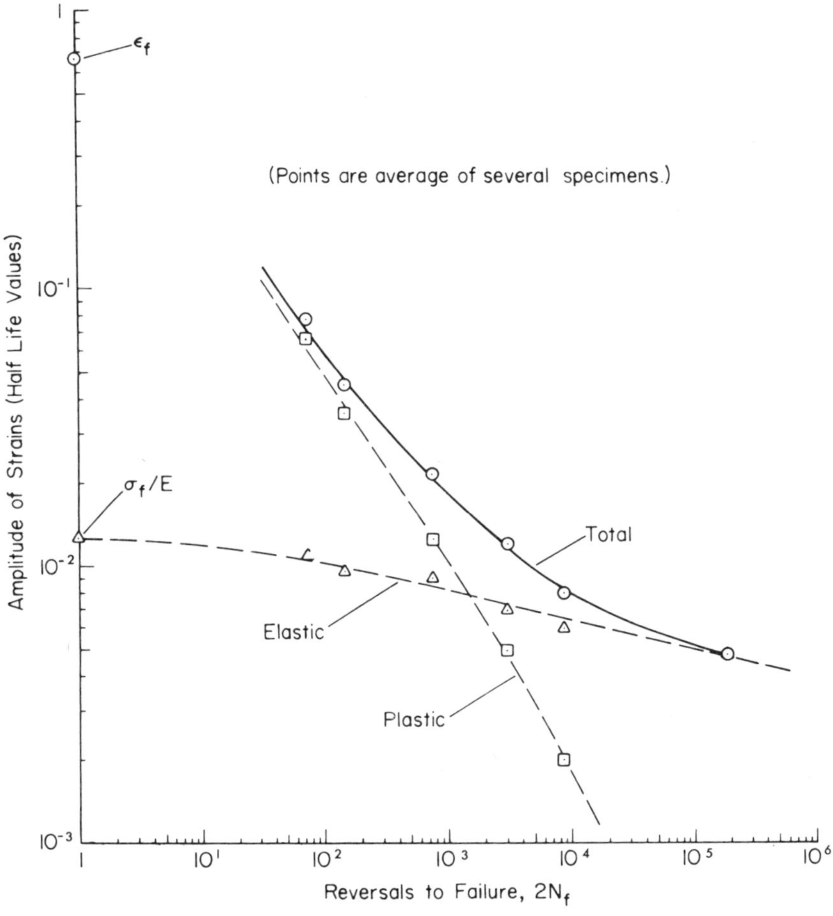

Fig. 58. Fatigue properties of 7075-T6 aluminum alloy (from Endo and Morrow, "Cyclic Stress-Strain and Fatigue Behavior of Representative Aircraft Metals." Courtesy ASTM)

The straight lines in Figs. 56 and 57 can be described by power functions whose general forms are of considerable universal significance in fatigue. The plastic line is given by

$$\frac{\Delta\epsilon_p}{2} = \epsilon_f'(2N_f)^c \quad \text{or} \quad 2\,N_f = \left(\frac{\Delta\epsilon_p}{2\epsilon_f'}\right)^{1/c} \tag{15}$$

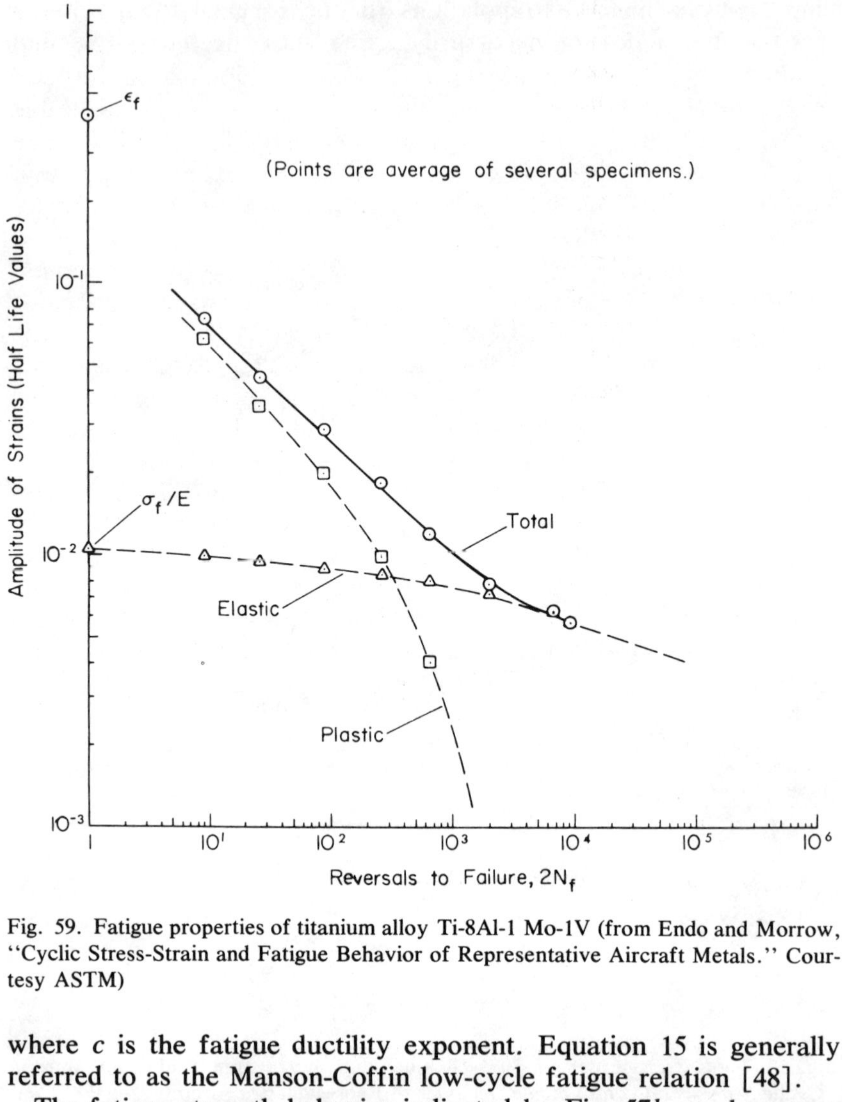

Fig. 59. Fatigue properties of titanium alloy Ti-8Al-1 Mo-1V (from Endo and Morrow, "Cyclic Stress-Strain and Fatigue Behavior of Representative Aircraft Metals." Courtesy ASTM)

where c is the fatigue ductility exponent. Equation 15 is generally referred to as the Manson-Coffin low-cycle fatigue relation [48].

The fatigue strength behavior indicated by Fig. 57*b* can be represented by a similarly simple equation,

$$\sigma_a = \sigma_f'(2N_f)^b \quad \text{or} \quad 2N_f = \left(\frac{\sigma_a}{\sigma_f'}\right)^{1/b} \tag{16}$$

which is obtained by combining Eq. 15 and the plastic strain component from Eq. 14. The resulting exponent cn' is called the fatigue strength exponent, b, which is often referred to as Basquin's exponent [48]. This is a small number which can cause a significant amplification of any error in σ_a. Equations 15 and 16 have only empirical bases at this point, but they will be discussed again on more physical grounds in the section on energy considerations.

It is necessary to elaborate on the definition of a fatigue coefficient such as that for ductility. This coefficient (ϵ_f') is normally taken as the intercept of the straight line in Fig. 57*a*, with the strain axis drawn at either 1/4, 1/2, or 1 cycle [34]. The 1/4 and 1/2 cycle can be called 1 reversal, and 1 cycle is 2 reversals. The difference between these possible locations of the strain axis may appear trivial at first sight, but the accurate prediction of the fatigue behavior from simple monotonic tests does depend on such ideas.

The above differences result from attempts to describe the monotonic tests in terms compatible with cyclic tests. In this respect it is helpful to use reversals (of σ or ϵ changing direction) instead of cycles. These are the easiest to count, within the first one or two cycles. On a reversal basis the monotonic tension test should be considered as one reversal, as in Fig. 57.

Another way to find the intercept is to consider the plastic strains in monotonic and cyclic deformations. We must remember that $\Delta\epsilon_p$ is a descriptor of the loop, just as $\Delta\sigma$ and $\Delta\epsilon$ are, and it is not the actual total plastic strain in a complete cycle, whereas ϵ_f is the actual plastic strain to fracture in monotonic tension. Thus it may be argued that the intercept ϵ_f' (at one reversal) of $\Delta\epsilon_p/2$ vs. $2N_f$ should be at $\epsilon_f/2$. In practice, this prediction is seldom found to be justified. Fig. 60 shows that $\epsilon_f' = \frac{1}{2}\epsilon_f$ is nearly always conservative and that $\epsilon_f' = \epsilon_f$ is best to represent the central trend. All of the above shows that accurate, generally applicable

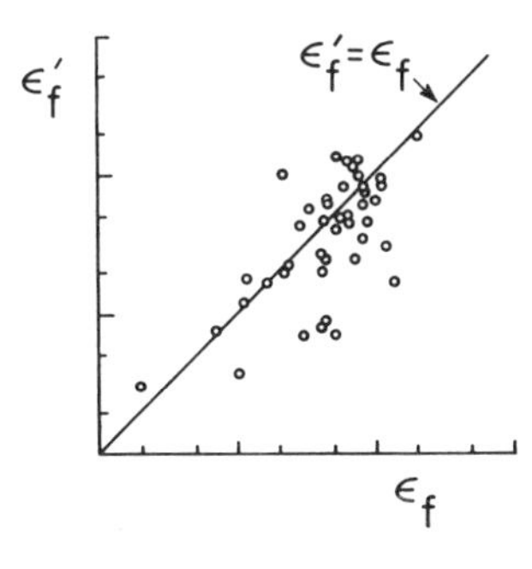

Fig. 60. Correlation between the fatigue ductility coefficient and the monotonic fracture ductility (after Landgraf, *High Fatigue Resistance in Metals and Alloys*)

methods for the prediction of the coefficients could not be expected. Unless otherwise noted, $\epsilon_f' = \epsilon_f$ will be used in this text, and $\epsilon_f = \Delta\epsilon_p/2$ will be plotted at one reversal.

Next let us explore the meaning of the fatigue ductility exponent. The main question concerns the damaging effect of cyclic plastic deformation relative to that of monotonic plasticity. This will be examined with the aid of the "exhaustion of ductility" model.

In a tension test, the maximum ductility available is the plastic strain ϵ_f. This is true regardless of any number of unloadings (Fig. 61) during the test, except when strain-aging occurs during the "rest"

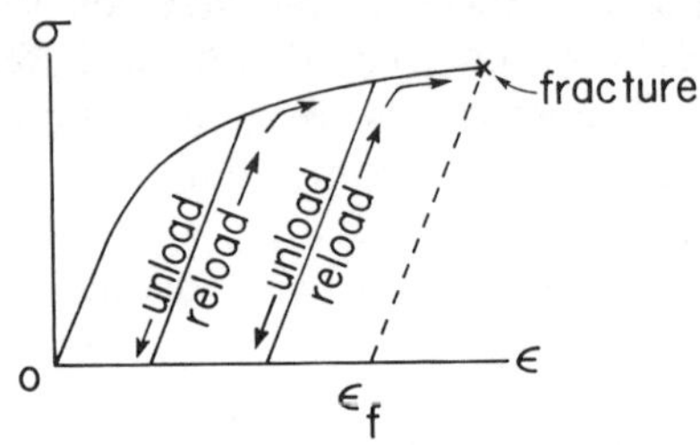

Fig. 61. Exhaustion of ductility in tension

period [11]. The question is whether the summation of plastic strains over all cycles in a fatigue test is also equal to ϵ_f. In other words, is ϵ_f a good measure of the total to-and-fro plastic flow that leads to a cyclic fracture in the same material? Superficially, only the tensile direction of the plastic flow appears to be damaging, since that is what finally leads to tearing the material apart. There is evidence, however, that cracks can initiate under shear stresses during compressive deformation [72]. This is significant in fatigue because plastic flow is caused only by shear stresses, which are present under both tensile and compressive loading.

The exhaustion of ductility concept can be discussed with the aid of Fig. 62. A bar of original length L is anchored at one end, and it is pulled to the right to fracture at L_1. The true fracture ductility is ϵ_f. An identical bar is pulled only to L_2 where the

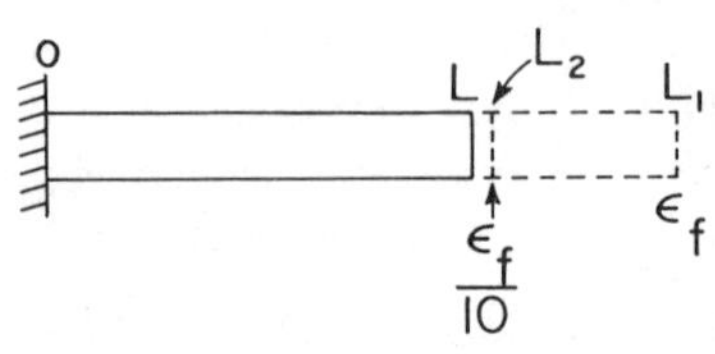

Fig. 62. Model for exhaustion of ductility

plastic strain is $\epsilon_f/10$. This bar is then pushed back to its original length L. In this complete cycle of deformation the material experiences a total plastic strain of $\epsilon_f/5$. Thus, on the basis of a primitive assumption that tensile and compressive deformations are equivalent, one-fifth of the material's ductility has been exhausted by the single cycle of loading. Under this magnitude of control condition the material can be expected to endure 5 cycles. After that, all of the ductility is exhausted and the bar should fracture. Similarly, if the plastic strain L_2 in Fig. 62 is $\epsilon_f/5000$, the ductility exhausted in a complete cycle is twice this much, and the life should be 2500 cycles. In units of reversals the fraction of ϵ_f describes the number of reversals the material should endure, and an inverse relationship results. Figure 63 shows such a relationship with a

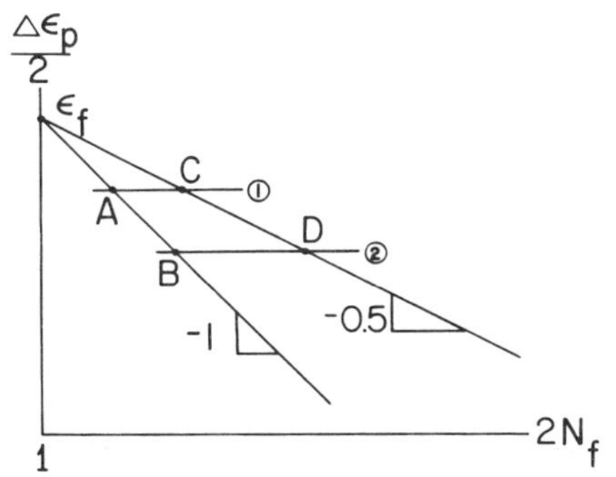

Fig. 63. Simple schemes of cyclic plastic strain vs. life

slope of -1, which indicates that the total plastic strain to fracture is a constant, regardless of the length of life. The formal way to state this is

$$\epsilon_f = \frac{\Delta\epsilon_p}{2}\,(2N_f) = \text{constant} \tag{17}$$

in which ϵ_f is monotonic, $\Delta\epsilon_p/2$ is fatigue, and $2N_f = 1$ is also monotonic. According to Eq. 17, a doubling of a life N_1 (at $\Delta\epsilon_{p,1}$), for example, is possible by halving the plastic strain per cycle. Real life, however, is again somewhat more complex than the simple theory. In low-cycle fatigue, most metals appear to have slopes of around -0.5 to -0.7 on a logarithmic plot of plastic strain vs. life. The range of slopes at long lives is not yet known. For an example, let us use -0.5 for the actual slope, as in Fig. 63. This means that if we go from the strain level 1 to half its value at level 2, the life is not doubled (from A to B) but quadrupled (in moving from C to D). If sufficient data are not available, a slope of -0.6 may be assumed for purposes of simple life estimation.

The meaning of the fatigue ductility exponent is that it reflects the material's tolerance to cyclic plastic strain without failure. A value of c smaller than -1 shows a tolerance greater than in monotonic tension. Another way to look at this is by attributing less damaging effect (or none for $c = -0.5$) to the compressive part of the cycle than to the tensile part. One could also say that as far as fatigue is concerned, ductility is not exhausted at equal rates in tension and compression. This is surprising, in view of the fact that shear stresses and consequently plastic strains result from compressive loading as well as from tensile. The possible initiation of cracks under compression further complicates the understanding of this peculiar behavior.

One conclusion that can be drawn from this is that it is not enough to initiate cracks; they also have to propagate and grow in size. Thus cracks may initiate in compression, and certainly in tension, but they will grow only in tension which opens the cracks and makes the stress concentrations at the crack tips a reality. In compression, the crack faces may close and temporarily reduce the stress concentration, and perhaps even cold-weld the flaw. It is believed that in fatigue there is a repeated process of ravaging the material, mainly in the tension part of a cycle. A chance for healing the damage is provided under compression. Of course, the destructive influences generally win the race.

The ductility exhaustion concept is quite simple, but it is by no means the only or the best explanation of damage under cyclic loading. The micromechanisms involved in crack initiation are surveyed in the works of Thompson and Wadsworth [73], Avery and Backofen [2], and Grosskreutz [23]. Paris and Erdogan show that crack propagation rates are related to the stress intensity factor [57]. It is generally recognized, as summarized by Laird [33], that the propagation of fatigue cracks often consists of two distinct stages. In Stage I the crack is oriented at approximately 45 degrees to the stress axis, and extends at a very low rate and only for a short distance. Stage II propagation is essentially perpendicular to the stress axis and is responsible for most of the actual fracturing of the material. It is believed that plastic deformations are the governing factors in both of these stages [33]. The continuum mechanics analyses summarized by Rice [61] provide various approaches to investigating the role of plasticity in the extension of fatigue cracks, and show the need for more research in this area.

3.2 The Transition Fatigue Life

Let us again consider the total strain vs. life curve as a summation of elastic strain vs. life and plastic strain vs. life curves (Eqs. 15, 16). The schematic representation of these is shown in Fig. 56. At short lives plastic strain obviously contributes more to the total strain than elastic strain does, and the reverse is true at long lives. There is a life where the two strain components are equal, as shown by the intersection of the straight lines. This life is called the transition fatigue life and is denoted by N_T or $2N_T$, depending on whether cycles or reversals are used. The transition life clearly marks the life region where each strain component dominates, but it should be remembered that even far away from N_T the smaller of the two components has a finite value.

The concept of the transition fatigue life has a great significance in the unified approach to the fatigue problem. Figure 64 shows

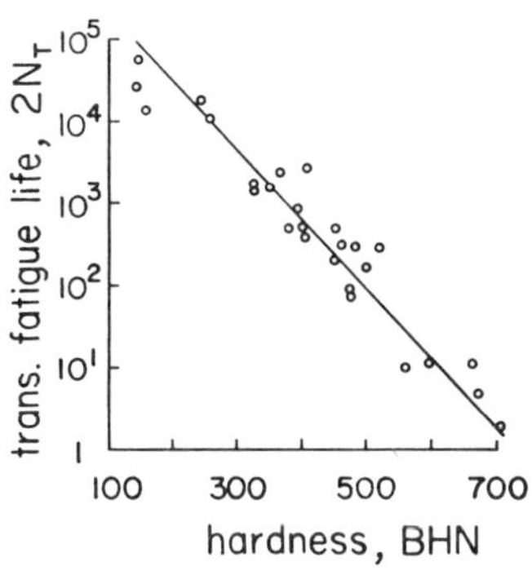

Fig. 64. Transition fatigue life as a function of hardness for steels (after Landgraf, *High Fatigue Resistance in Metals and Alloys*)

the orders of magnitude variation in transition life for steels of various hardness. This life can be at only a few cycles for the hardest steels, but it can be close to 10^5 reversals for the most common steels. The latter means considerable plastic strain at 50,000 cycles, which is about the highest of the various proposed upper limits to the low-cycle realm. A transition life of 10^5 reversals also means that, for example, even at 10^6 reversals to failure there is considerably more than zero plastic strain per cycle present, and that it would be foolish to think strictly in terms of long-life fatigue, hoping that low-cycle concepts would not be applicable.

Transition fatigue lives should receive increasing attention in the future, especially in engineering design and material selection.

3.3 The Desirable Life Curve

One could ask whether the σ-N_f (or S-N) or the ϵ-N_f curve is the more useful, or perhaps if both of them are needed to cover the large range of lives at which a material may have to serve. The most useful and most desirable curve is an extended total strain vs. life plot covering a life range of about 10^7 cycles. Such a curve is more difficult to obtain than either the S-N for long life or the ϵ-N_f for short life, but it presents the most information in a single diagram when the elastic and plastic lives are also shown. For example, it is useful to recognize that an ϵ-N_f curve has the S-N curve locked up in it in the form of the elastic strain (σ/E) vs. life line. The ϵ-N_f plot thus presents a universally desirable picture of a material's fatigue behavior.

It must be admitted that an extended ϵ-N_f curve does present some complications. The main difficulty is that at long lives even the total strain becomes quite small, and the control of these small, cyclically varying quantities requires accurate instrumentation and extreme precision in test procedures. A temporary solution of this problem is to use stress control at long lives, and plot the calculated elastic strain vs. life on the total strain vs. life diagram (Fig. 65).

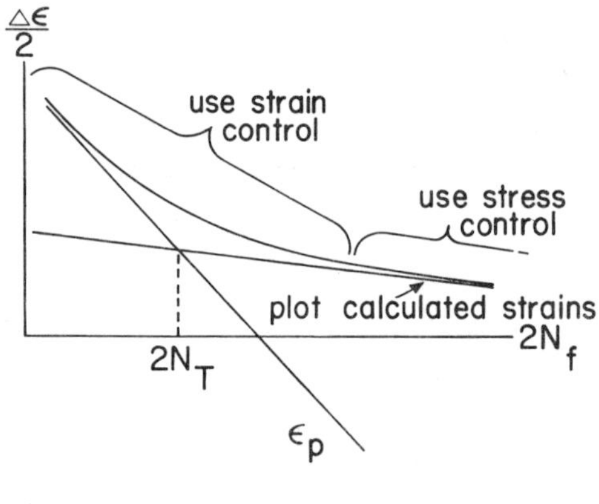

Fig. 65. Practical considerations concerning the most desirable life curve

In this life range the plastic strain component is so small (continue log plot below horizontal axis) that the ϵ-N_f curve appears the same for both ϵ and ϵ_e.

The value of the ϵ-N_f curve depends on the recording of hysteresis loops throughout the life of each specimen tested. It is obviously not a good idea, however, to record every single loop, even at a life of 1000 or 100, because the quantity of data would soon become unmanageable. The method recommended is to take advantage of the exponential nature of such fatigue processes as cycle-dependent adjustments caused by softening or hardening. A simple and economical procedure is to take hysteresis loops in the following order of cycles: 1st, 2nd, 5th, 10th, 20th, 50th, 100th, 200th, 500th, 1000th, 2000th, 5000th, and so on. As an example, consider a specimen that failed at 71,000 cycles and whose loops at half-life and at failure are required. Neither one of these loops is available exactly, but we can closely approximate them. The last loop taken was at 50,000 cycles, and any cycle-dependent change in the last 21,000 cycles would be negligible. Similarly, the loop for half-life lies between those taken at 20,000 and 50,000 cycles, and the difference between these two available loops should be relatively small.

4

Energy Considerations in Fatigue

Energy is dissipated in fatigue because of plastic deformation. Most of this energy is converted into heat and is not recoverable as strain energy. The energy per cycle, ΔW, is measured by the area of the hysteresis loop, and the total energy expended during the life is a summation of all of the loop areas. In those cases where ΔW does not vary much from cycle to cycle in a given test, the total plastic strain energy, W_p, is described by the equation

$$W_p = \Delta W N_f \tag{18}$$

W_p is called the fatigue toughness [25]. A striking property of the fatigue toughness is that it tends to increase, for a given material, with the number of cycles to failure. The following inequality is a simple statement of this observation:

$$W_f \text{ in monotonic loading} < W_p \text{ in low-cycle fatigue} < W_p \text{ in high-cycle fatigue} \tag{19}$$

Such a general tendency is seen in Fig. 66, which shows the total energy to fracture as a function of life for more than 1200 individual tests on metals under various control conditions [25].

It is interesting to note that from the energy point of view fatigue is not a bad problem. Normally, fatigue is considered as a seriously weakening process causing fractures at stresses below the ultimate or even the yield strengths. At the same time it is possible to

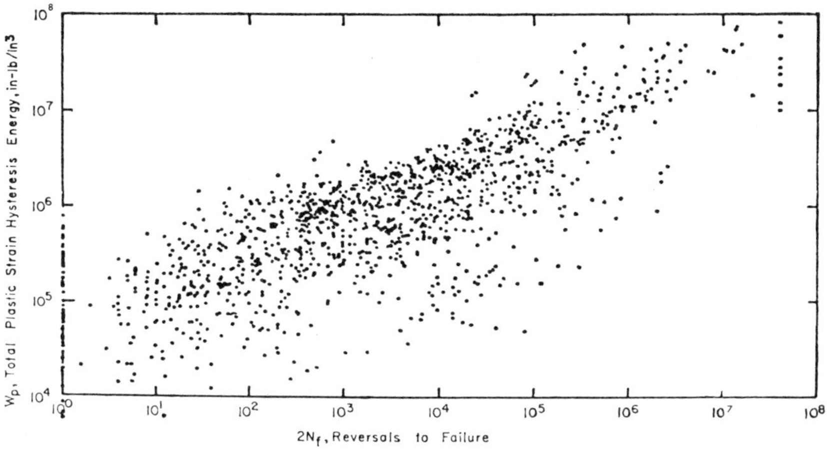

Fig. 66. Fatigue toughness as a function of life; about 1200 fatigue test results are included (from Halford, "The Energy Required for Fatigue." Courtesy ASTM)

find that 100 times the monotonic fracture energy may be required to cause failure in half a million cycles [25].

4.1 Failure Prediction

Hysteresis energy is a useful basis for the establishment of failure criteria in fatigue [18, 44, 61]. Energy considerations in this text follow the simple concept [48] that elastic strain energy is applicable mainly to the last cycle of fatigue, and plastic strain energy is useful in predicting the accumulation of damage before final failure.

The last cycle is important because that is where the final separation occurs. For brittle materials the fracture criterion is given by the Griffith criterion [72],

$$\sigma_F = \sqrt{\frac{2E\gamma}{\pi L}} \quad \text{or} \quad \frac{\sigma_F^{\,2}}{2E} L = \text{constant} \tag{20}$$

where σ_F = nominal fracture stress, γ = surface energy, and L = crack length.

In words, Eq. 20 means that a crack becomes self-propagating when the decrease in elastic strain energy (by unloading upon fracture) is at least equal to the energy required to create the new crack surfaces.

In fully reversed cycling the maximum elastic strain energy is

$$\frac{\sigma_a^{\,2}}{2E} = \frac{1}{2}\,(\sigma_a \cdot \epsilon_e) \tag{21}$$

Thus for two different amplitudes, σ_1 and σ_2, the critical crack lengths for final fracture are given on the basis of Eq. 20* as

$$\frac{L_1}{L_2} = \left(\frac{\sigma_2}{\sigma_1}\right)^2 \tag{22}$$

The curves shown earlier in Fig. 52 also represent the behavior expressed in Eq. 22. Final fracture in fatigue occurs with the prior development of large cracks when the stresses are small, and vice versa.

Equation 20 is the fracture criterion both in fatigue and in monotonic loading, but there is an essential difference between the two cases in reaching the point of fracture. In monotonic loading, the stress is gradually increased by raising the external load until fracture occurs. In fatigue, the stresses increase because the gradually growing cracks reduce the load-bearing area. This may not be entirely clear when a stress control situation is intended. We must remember that true stress control is all but impossible when cracks are present, and even this ideal control condition eventually degenerates to a simple load control.

The explanation above shows that crack growth can create the critical stress to cause final fracture. In the following discussion, cyclic plastic strain energy is used as a criterion for crack propagation. The analysis is not exact, in order to allow a simple treatment of the complex fatigue problem.

A cracked material under load will deform plastically first at the notch tips, as in Fig. 67. The size of a plastic zone is loosely

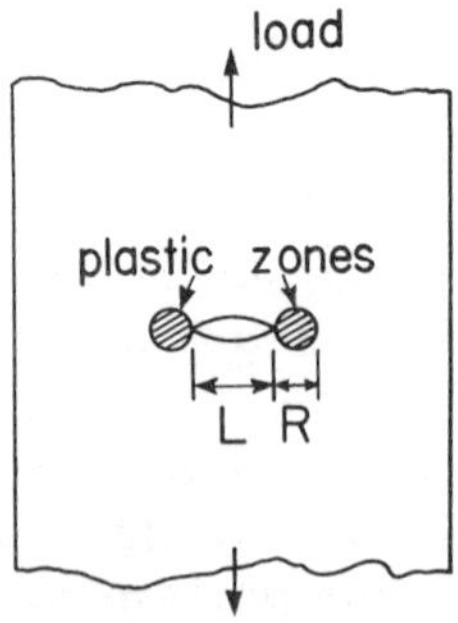

Fig. 67. Model of plastically deformed zones at crack tips

*The criterion can be modified to account for plasticity. The relation for stress intensity factor used in fracture mechanics [72] is also applicable.

called the radius R of the plastic zone, even though it is not necessarily the radius of a circle. In general, R is believed to be linearly dependent on the crack length [72]. The energy required to plastically deform the material near the crack depends on the volume of the plastic zone and on the strain distribution within the zone. This volume is a function of the crack size, as R itself (essentially a one- or two-dimensional concept) depends on the crack length. It is not clear what this function should be to deal with the fatigue problem correctly. Two possibilities will be discussed to show the simplest methods of handling the problem.

First, let us assume that the plastically deformed volume associated with a crack is proportional to the cube of the crack length. We also assume that, regardless of the exact mechanism, in this plastically deformed zone the condition is met to propagate the crack. Thus, the plastic strain energies required to propagate two different sizes of cracks are related to each other by

$$\frac{W_{p_1}}{W_{p_2}} = \left(\frac{L_1}{L_2}\right)^3 \tag{23}$$

Using Eq. 22, stresses and energies can be retained while crack lengths are eliminated,

$$\frac{\sigma_2}{\sigma_1} = \left(\frac{W_{p_2}}{W_{p_1}}\right)^{-1/6} \tag{24}$$

In Eq. 24 we can use monotonic fracture values for σ_1 and $W_{p,1}$ and fatigue values for σ_2 and $W_{p,2}$,

$$\frac{\sigma_a}{\sigma_f} = \left(\frac{W_p}{W_f}\right)^{-1/6} \tag{25}$$

where σ_a = cyclic stress amplitude, σ_f = true fracture stress, monotonic, W_p = total cyclic plastic work or fatigue toughness at σ_a, and W_f = true toughness, monotonic.

For another solution, let us assume that the plastic volume at a crack tip is proportional to the square of the crack length. Through a derivation identical with the preceding one (Eqs. 22, 23, 24) the counterpart of Eq. 25 becomes

$$\frac{\sigma_a}{\sigma_f} = \left(\frac{W_p}{W_f}\right)^{-1/4} \tag{26}$$

Either of the two exponents may give good results for some metals, but the exponent $-1/4$ is the most likely to be found in the fatigue literature [48]. Equations 25 and 26, or similar ones, can be checked by plotting σ_a against W_p on log-log coordinates. The line should go through a point for W_f and σ_f and have a slope equal to the exponent of the correct equation. It is found [48] that for most metals the slopes are between $-1/8$ and $-1/3$. This makes either Eq. 25 or 26 admissible until more precise relationships become available.

The above results can lead to a relationship between stress amplitude or cyclic plastic strain and fatigue life. To obtain a formal expression it is necessary to describe the plastic strain energy in each cycle mathematically.

It is best to deal directly with stress vs. plastic strain quantities as has been done with the loop in Fig. 68. The loop was made

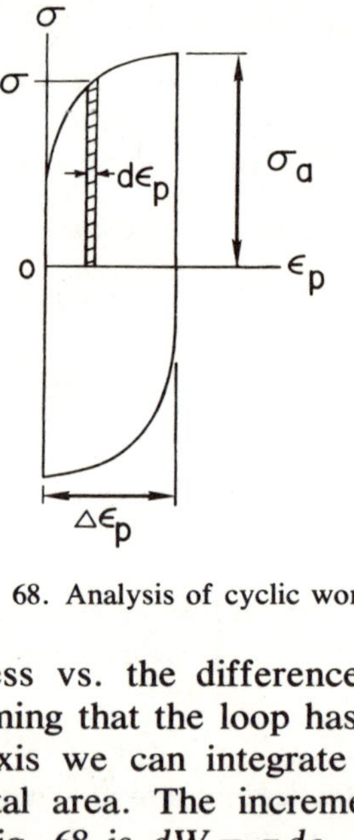

Fig. 68. Analysis of cyclic work

vertical by plotting stress vs. the difference between total strain and elastic strain. Assuming that the loop has identical areas above and below the strain axis we can integrate in the first quadrant and obtain half the total area. The incremental work shown by the shaded section in Fig. 68 is $dW = \sigma\, d\epsilon_p$. The total area inside the loop is the hysteresis energy per cycle,

$$\Delta W = 2\int_0^{\Delta\epsilon_p} \sigma\, d\epsilon_p \tag{27}$$

The nonlinear part of the loop can be expressed by a power function such as that for monotonic deformation,

$$\sigma = K \epsilon_p^n \tag{28}$$

Substituting and integrating,

$$\Delta W = \frac{2K}{n+1} (\Delta\epsilon_p)^{n+1} = \frac{2}{n+1} (\Delta\epsilon_p) [K(\Delta\epsilon_p)^n] \tag{29}$$

where $K(\Delta\epsilon_p)^n = \sigma_a$ according to Fig. 68. Therefore, the plastic work per cycle is

$$\Delta W = \frac{2}{n+1} \sigma_a(\Delta\epsilon_p) \tag{30}$$

Equation 30 is a linear relationship even though the stress-strain curve itself is nonlinear. This is a property of all areas under curves expressed by power functions.

For the present analysis we assume that Eq. 30 represents the average plastic strain energy per cycle, or that no changes occur from cycle to cycle. The total work during fatigue life is given by Eq. 18,

$$W_p = \Delta W N_f$$

Substituting Eqs. 26 and 30,

$$N_f \left(\frac{2}{n+1}\right) \sigma_a (\Delta\epsilon_p) = W_p = W_f \left(\frac{\sigma_a}{\sigma_f}\right)^{-4} \tag{31}$$

Before continuing the derivation let us consider the power function relating stress and plastic strain as explained for Eq. 13,

$$\epsilon_p = \left(\frac{\sigma}{K}\right)^{1/n} = \epsilon_f \left(\frac{\sigma}{\sigma_f}\right)^{1/n}$$

Under cyclic loading a single application of the stress amplitude σ_a will produce the so-called plastic strain range $\Delta\epsilon_p$ (as in Fig. 68). Therefore

$$\Delta\epsilon_p = \epsilon_f \left(\frac{\sigma_a}{\sigma_f}\right)^{1/n} \tag{32}$$

and

$$\frac{\sigma_a}{\sigma_f} = \left(\frac{\Delta\epsilon_p}{\epsilon_f}\right)^n \tag{33}$$

Note that a blindly used terminology of stress amplitude σ_a and associated plastic strain amplitude $\Delta\epsilon_p/2$ now incorrectly gives

$$\frac{\Delta\epsilon_p}{2} = \epsilon_f\left(\frac{\sigma_a}{\sigma_f}\right)^{1/n} \quad \text{and} \quad \frac{\sigma_a}{\sigma_f} = \left(\frac{\Delta\epsilon_p}{2\epsilon_f}\right)^n \tag{34}$$

However, the difference between Eqs. 33 and 34 is not large, since a likely range of $0.1 < n < 0.3$ makes $2^n \approx 1$.

Let us now return to Eq. 31. The monotonic energy to fracture, W_f, can be expressed in much the same way as the cyclic strain energy obtained in Eq. 30, except that the reversal in load is not included, so that

$$W_f = \frac{1}{n+1}\sigma_f \epsilon_f \tag{35}$$

Substituting in Eq. 31,

$$N_f\left(\frac{2}{n+1}\right)\sigma_a(\Delta\epsilon_p) = \left(\frac{1}{n+1}\right)\sigma_f \epsilon_f \left(\frac{\Delta\epsilon_p}{\epsilon_f}\right)^{-4n} \tag{36}$$

After multiplying by $(n+1)$ and rearranging,

$$(2N_f)^{1/(1+4n)}(\sigma_a)^{1/(1+4n)}\left(\frac{\Delta\epsilon_p}{\epsilon_f}\right) = (\sigma_f)^{1/(1+4n)} \tag{37}$$

Using Eq. 33 we can eliminate $\Delta\epsilon_p/\epsilon_f$, so that

$$(2N_f)^{n/(1+4n)}(\sigma_a)^{[n/(1+4n)]+1} = (\sigma_f)^{[n/(1+4n)]+1}$$

which can be simplified to

$$\sigma_a(2N_f)^{n/(1+5n)} = \sigma_f \tag{38}$$

or

$$\log(\sigma_a) = \log(\sigma_f) - \frac{n}{1+5n}\log(2N_f)$$

where $-n/(1+5n) = b$. In Eq. 37 we could retain the strain term and eliminate σ_a and σ_f to give

$$(\Delta\epsilon_p)(2N_f)^{1/(1+5n)} = \epsilon_f \tag{39}$$

or

$$\log(\Delta\epsilon_p) = \log(\epsilon_f) - \frac{1}{1+5n}\log(2N_f)$$

where $-1/(1+5n) = c$. This equation is best at short lives where plastic strains are known.

Equations 38 and 39 indicate that we can predict the fatigue life for a specified control condition σ_a or $\Delta\epsilon_p$ on the basis of monotonic properties, the strain-hardening exponent (n), and the fracture coefficients (ϵ_f, σ_f). These predictions are quite crude if considerable cycle-dependent adjustment in stress-strain response takes place. Therefore the equations can be refined if the correct value of n' is used instead of n. In fact, an educated guess can be made even if n' is not known precisely. This is done on the basis of general tendencies in changes of n to n'. For example, if n is above the stable range of 0.1 to 0.2, 0.42 for instance, we could expect much cycle-dependent hardening. In this case n' should be smaller than n, but not below the stable range, so a value of 0.2 may be a reasonable guess for use in either Eq. 38 or 39.

A word of caution is in order concerning the use of Eqs. 38 and 39. The accuracy of the results cannot always be guaranteed, as the following few examples show. The logarithmic life plots may make the magnitude of the deviations not very obvious, however. In any case, it is often possible to get into a desired life range by the use of the above crude methods of prediction, and save a number of tests in the process.

4.2 Examples

Example 1

7075-T6 Aluminum alloy. See [14]
$\epsilon_f = 0.41$ $\qquad \epsilon_f' \approx 0.18$

$\sigma_f = 108$ ksi (corrected for necking) $\sigma_f{}' \approx \sigma_f$
$n = 0.113$ $n' = 0.11$
At $\sigma_a = 73$ ksi, $\Delta\epsilon_p = 0.0013$ (stable values)
What is the fatigue life?
From Eq. 39,

$$2N_f = \left(\frac{\epsilon_f}{\Delta\epsilon_p}\right)^{1+5n} = \left(\frac{0.18}{0.0013}\right)^{1.55} = 2000 \text{ reversals}$$

This is the correct result found by experiments in [14]. From Eq. 38,

$$2N_f = \left(\frac{\sigma_f}{\sigma_a}\right)^{(1+5n)/n} = \left(\frac{108}{73}\right)^{14.1} = 260 \text{ reversals}$$

which falls too far short of the actual value of 2000.

Example 2

2024-T4 Aluminum alloy. See [14].
$\epsilon_f = 0.43$ $\epsilon_f{}' \approx 0.18$
$\sigma_f = 91.5$ (corrected for necking) $\sigma_f{}' \approx \sigma_f$
$n = 0.2$ $n' = 0.11$
At $\sigma_a = 65$ ksi, $\Delta\epsilon_p = 0.0039$ (stable values)
From Eq. 39,

$$2N_f = \left(\frac{0.18}{0.0039}\right)^{1.55} = 380 \text{ reversals}$$

which falls too far short of the actual 2000 reversals. Interestingly enough, using ϵ_f instead of $\epsilon_f{}'$ would give

$$2N_f = \left(\frac{0.43}{0.0039}\right)^{1.55} = 1500 \text{ reversals}$$

which is fairly close to the actual number.

Example 3

2024-T4 aluminum alloy. See [14]
Data are same as in Example 2
At $\sigma_a = 57$ ksi, $\Delta\epsilon_p = 0.0003$ (stable values)
From Eq. 39,

$$2N_f = \left(\frac{0.18}{0.0003}\right)^{1.55} = 20{,}000 \text{ reversals}$$

which is to be compared with the actual life of 8400 reversals. The most likely source of error in this example is the small value of $\Delta\epsilon_p$, which is about as small as can be conveniently measured. Such a minute quantity is difficult to measure or control accurately.

Example 4

SAE 4340 steel, hard condition. See Fig. 57 and Ref. 48
$\sigma_f = \sigma_f' = 174$ ksi
$n' = 0.15$

What is the fatigue life at $\sigma_a = 50$ ksi? From Eq. 38,

$$2N_f = \left(\frac{\sigma_f}{\sigma_a}\right)^{(1+5n)/n} = \left(\frac{174}{50}\right)^{11.7} = 2.1 \times 10^6 \text{ reversals}$$

which is approximately correct, according to Fig. 57.

The examples above show that fatigue life predictions on the basis of energy considerations do not yet have a sufficient chance of success. It must be remembered that it is dangerous to rely on calculated values alone. There is a great need for continued research in the area of fatigue life predictions from simple monotonic and fatigue properties such as ϵ_f, σ_f, n, ϵ_f', σ_f', and n'.

5

Cumulative Damage

The discussion up to this point has dealt with the simplest possible problems in fatigue. Specimens considered had relatively uniform, smooth features, they were not tested in hostile environments, the applied loads were fully reversed in tension and compression, and the stress or strain amplitudes were constant throughout the life. There are only a few practical situations where all of these simplifying conditions are satisfied.

The special case discussed here is one in which the amplitude is not a constant. The most common of this sort is the random fatigue situation shown in Fig. 69. The problems illustrated arise

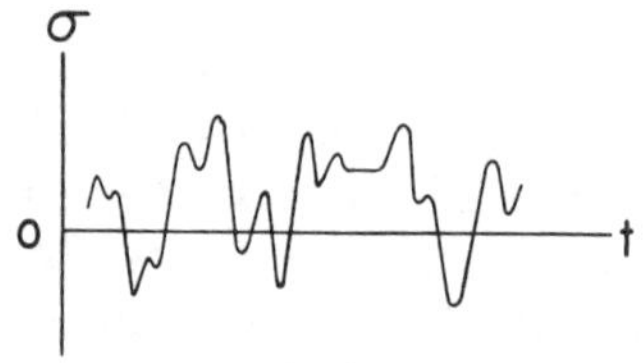

Fig. 69. Random loading

often in vehicle suspensions, for example. Special techniques have been developed to deal with the complex wave forms involved [24, 71], and they will not be discussed in detail here. Attention will be given, instead, to the relatively simpler problems associated with multiple-level loading, where the amplitude is a constant within blocks of cycles, as in Fig. 70. These conditions exist in machinery that is operated at discrete speeds. For the present purpose this approach is important in serving as a stepping stone between single-

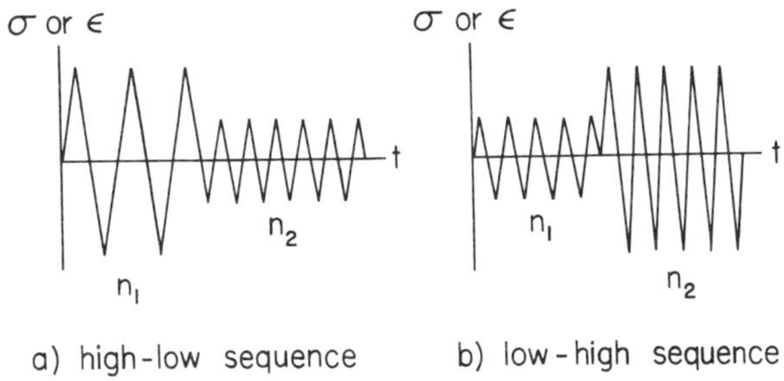

Fig. 70. Simple step-loading, showing high-to-low and low-to-high sequences

level fatigue and random loading. Actually, it is adequate in some cases to perform multiple-level block programs to simulate random service loads.

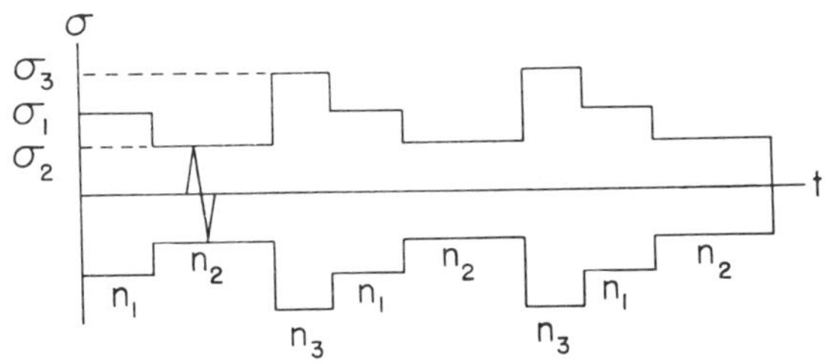

Fig. 71. Multiple-step programming

The terminology of step loading is shown in Figs. 70 and 71. The controlled quantity is either stress or strain, but in a given program it is generally one or the other without switching from block to block. At each given level of load the number of cycles is denoted by the same subscript as for the control function. In the simplest but highly useful case of two-level, two-block loading, the number of cycles in the second block is not predetermined, so that cycling may lead to failure in two blocks. This means that the number of blocks is specified. In a more general case, like that illustrated in Fig. 71, the sequence (or group) of blocks is repeated until fracture, so that the number of full sequences is itself a measure of the fatigue life. A complete sequence such as $n_1 + n_2 + n_3$ (Fig. 71) is often called a block, but here block will mean that there is only constant amplitude within it.

There are a number of special problems that arise in step loading and require individual attention. These include high-to-low (often written as Hi-Lo) or low-to-high (Lo-Hi) transfer, the possible generation of mean stresses, the relative load levels between blocks, and the number of cycles in a particular block. Another variable, the number of identical groups of blocks during life, is apparently not an important factor [76].

The material's character in responding to applied stresses and strains can be just as evident under stepwise loading as under steady amplitudes. To observe the response it is necessary to record the independent variable. Figure 72 shows schematically one possible

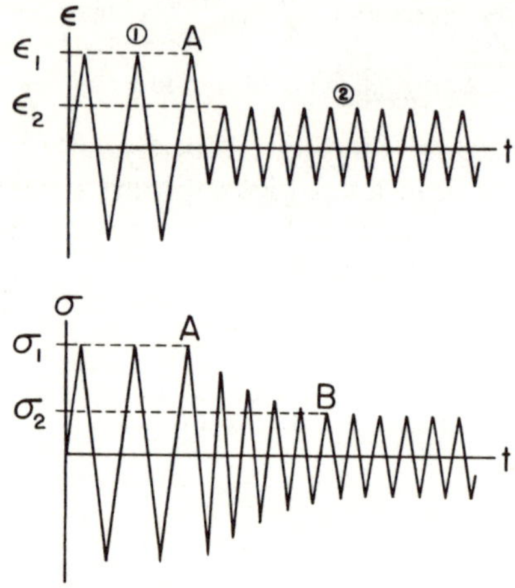

Fig. 72. Schematic of cycle-dependent adjustment of independent variable under step change of control function

situation. The control condition here is cyclic strain, and because this condition is externally enforced it is possible to step at once from ϵ_1 to ϵ_2. The stresses caused by the enforced strains in the material are σ_1 and σ_2, but the transition in stress is quite smooth compared to the abrupt change in strain. This adjustment is at an exponentially decreasing rate and is similar to cycle-dependent softening in this particular case. An explanation of the phenomenon can be found by considering block 1 and block 2 as independent, single-level tests. In these, if there is cycle-dependent adjustment, the material would reach the steady states of σ_1 and σ_2 only after a number of cycles. We could say, for example, that at ϵ_2 the natural resistance of the material is σ_2, but it is reached only gradually. At point *A*, also, the resistance to deformation represented by σ_1 has been reached over a number of cycles in block 1. This is true regardless of softening or hardening in that block, since only the end result at *A* counts now. Therefore even after a change to ϵ_2 (in Fig. 72) the material remembers (by its internal structure) that it had been cyclically worked generating stresses of σ_1, and it displays more strength than expected for a while.

The responses shown in Fig. 73 show data for four possible

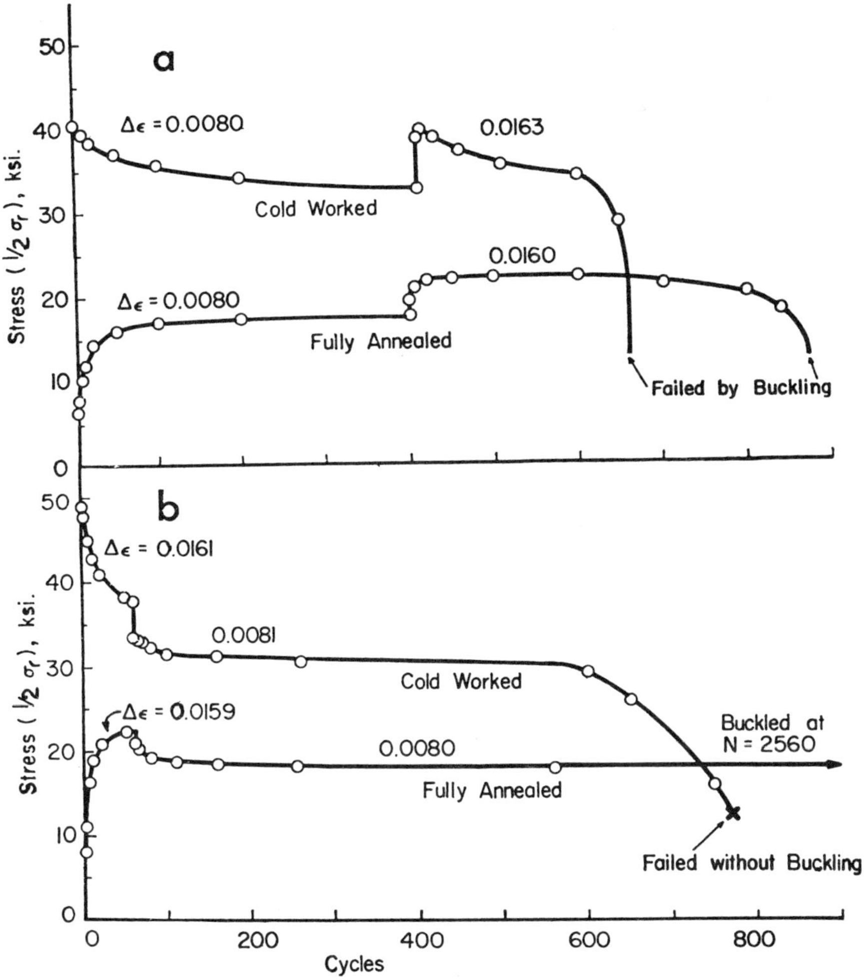

Fig. 73. Stress adjustments in copper under controlled cyclic strains: (a) low-to-high cyclic strain; (b) high-to-low cyclic strain (from Tuler, in Morrow, *Internal Friction, Damping, and Cyclic Plasticity*. Courtesy ASTM)

situations with low-to-high or high-to-low steps and softening or hardening materials. Part *b* of the figure illustrates the case described above. An opposite example is the annealed (hardening) metal in a low-to-high step, as shown in Fig. 73*a*. There is considerable hardening in the first block, but the metal still has a capacity for further cycle-dependent hardening after the step to higher strains. This hardening is similar to that in the beginning of the life, except that eventually the higher strains force the stress to a higher level.

The transient response to steps of stress or strain is generally not apparent in the final count of the fatigue life. It could become important, however, where the number of cycles in a block is small so that the independent variable never has a chance to stabilize during the block. In such a case only actual measurements of the independent variable can give its magnitude from cycle to cycle, and the life may be affected to a small extent by the transients.

It is of considerable practical interest to understand the step-loading process and be able to predict the resulting life. There are numerous theories [24] that have been developed to deal with the problems of "cumulative damage" arising from complex loading patterns. The term is often used incorrectly because it is generally implied that nonsteady amplitudes are involved. Of course, the essence of fatigue is that damage gradually accumulates from cycle to cycle, whether it is a life of a few cycles or of millions, and whether the loading is constant throughout the life or not.

5.1 The Exhaustion of Life

The fatigue process can be looked on either as an accumulation of damage toward a maximum tolerable damage, or as an exhaustion of the life (or ductility) inherent in the material. To analyze and predict the outcome of complex loading, let us first discuss the simplified schematic life diagram in Fig. 74. At a constant σ_1 the

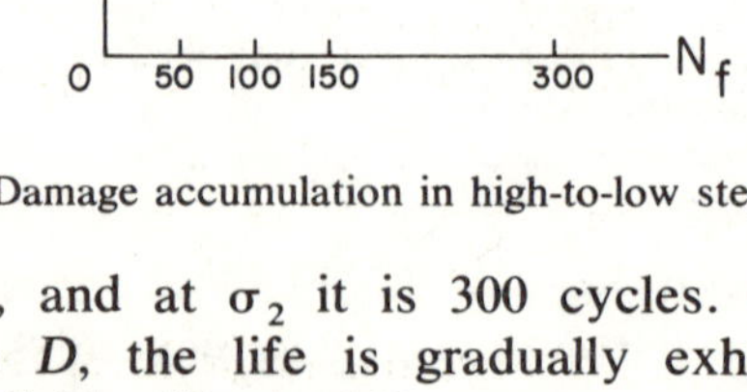

Fig. 74. Damage accumulation in high-to-low step loading

life is 150 cycles, and at σ_2 it is 300 cycles. Going from *A* to *B*, or from *C* to *D*, the life is gradually exhausted. At points *A* and *C* the available life is 100 percent at each level, and at *B* and *D* the lives have been spent.

If we assume a linear accumulation of damage, each cycle contributes the same amount of damage at a given level of stress. At σ_1 we may, for example, cycle from A to E and exhaust one-third of the life in the process. Consider that after reaching E the stress level is changed to σ_2. In analyzing the results of such a step we will assume that a certain percentage of life exhausted at one level of stress is equivalent to the exhaustion of the same percentage of life at any different stress amplitude. Thus one-third of the life at σ_1 is one-third of the life at σ_2, and in the step we go from point E to F at 100 cycles. Since only one-third of the life has been exhausted at σ_1 we expect two-thirds of it to be still available at σ_2 to give failure at point D, even though two separate levels of stress are involved.

A low-to-high step in Fig. 75 is basically the same as the high-to-

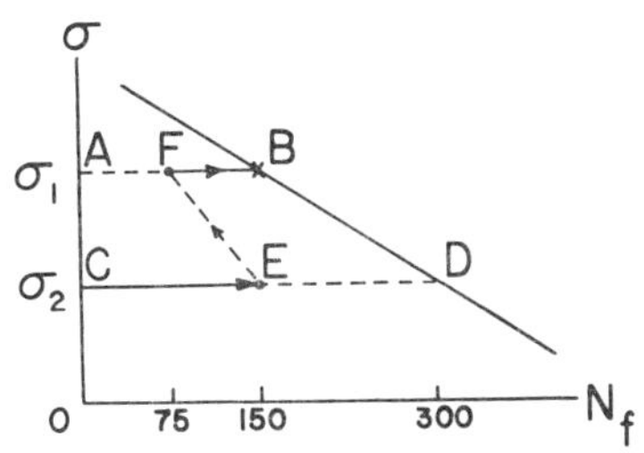

Fig. 75. Damage accumulation in low-to-high step loading

low change described above. For example, 150 cycles at σ_2 is 50 percent of the life. Stepping to σ_1 means that we have to start at 75 cycles, which is half the life at the new level. If the damage is linearly accumulated, fracture should occur at point B.

The models above show that, at least with certain assumptions, the fatigue life after a complex loading pattern may still be predicted if the complete life curve for single-level cycling is known. The method described is generally referred to as the Palmgren-Miner law, simply as Miner's rule, or as the linear cumulative damage theory [47, 56]. The formal statement of the linear damage concept is that the summation of all fractions of life equals one,

$$\sum \frac{n_i}{N_{fi}} = 1 \quad \text{or} \quad \frac{n_1}{N_1} + \frac{n_2}{N_2} + \frac{n_3}{N_3} + \ldots = 1 \tag{40}$$

where $N_1, N_2, N_3, \ldots N_{fi}$ are the single-level fatigue lives at given levels of loading, and $n_1, n_2, n_3, \ldots n_i$ are the actually performed numbers of cycles at the specified levels.

Example

2024-T4 Aluminum alloy has the following fatigue lives under strain control [14]:

at $\Delta\epsilon_1 = 0.058, \quad N_1 = 45$ cycles
$\Delta\epsilon_2 = 0.0245, \quad N_2 = 310$
$\Delta\epsilon_3 = 0.0093, \quad N_3 = 12{,}400$

How many times can the following sequence of blocks be repeated?

$n_1 = 5$ cycles at $\Delta\epsilon_1$
$n_2 = 60$ cycles at $\Delta\epsilon_2$
$n_3 = 495$ cycles at $\Delta\epsilon_3$

The fractions of life exhausted in each block are

$$\frac{n_1}{N_1} = \frac{5}{45} = 0.111, \quad \frac{n_2}{N_2} = 0.194, \quad \frac{n_3}{N_3} = 0.04$$

The fraction of life exhausted in a complete sequence of the three given blocks is approximately 0.345. According to Eq. 40 the life is entirely exhausted when the sequence of blocks is repeated X times; $X(0.345) = 1, \quad X = 2.9$. Thus it is likely that the failure will occur under strains of $\Delta\epsilon_2$ during the last block of the overall sequence $n_1 + n_2 + n_3 + n_1 + n_2 + n_3 + n_1 + n_2$.

It must be emphasized that the correct application of the linear damage rule depends on several strongly limiting assumptions. In each block the loading must be fully reversed (no mean stresses). The rate of damage accumulation at any given level must not depend on prior loading history. This means that the damage per cycle should be the same early and late in the life for each level. It also means that the magnitude of the amplitude change should be immaterial, and the order of the change (high-to-low or low-to-high) must not affect the life. Each one of these assumptions is potentially unrealistic. For example, the order of the change may have an effect (also called the sequencing effect). It may be found that for identical block sizes and amplitudes the high-to-low order is more damaging than the low-to-high one [39]. This is the case where cracks initiate during the larger loads and can continue growing at the following smaller loads, whereas the smaller loads themselves would not rapidly initiate the cracks. Other aspects of these complicated situations in cumulative damage will be discussed in later sections.

When the most important assumptions are fairly well satisfied, the linear damage calculations are quite reliable, as shown by data in Table 2. For perfect results, each of the damage summations

Table 2. Strain-controlled step tests on 2024-T4 aluminum alloy; no mean stresses

No. of blocks and sequence	$\Delta\epsilon_1/2$	n_1 per block	Mean stress, ksi	$\Delta\epsilon_2/2$	n_2 per block	Mean stress, ksi	Damage $\Sigma n_1/N_{fi}$
1 hi-lo	0.0512	4	0	0.0116	202	0	0.92
1 lo-hi	0.0124	175	0	0.0518	9	0	1.38
1 hi-lo	0.0181	60	0	0.0070	1600	0	1.10
1 lo-hi	0.0070	850	0	0.0181	105	0	1.09
4 hi-lo	0.0526	2	0	0.0124	70	0	1.26
4 lo-hi	0.0125	70	0	0.0527	2	0	1.32
5 hi-lo	0.0182	13	0	0.0072	230	0	1.00
5 lo-hi	0.0072	230	0	0.0182	13	0	0.98
3 hi-lo	0.0106	100	0	0.0036	2000	0	0.98
3 lo-hi	0.0034	10 000	0	0.0100	100	0	1.00
7 hi-lo	0.0099	50	0	0.0050	850	0	1.24
5 lo-hi	0.0048	850	0	0.0099	50	0	0.83
11 lo-hi	0.0150	17	0	0.0300	1	0	1.13
21 lo-hi	0.0114	17	0	0.0224	1	0	1.19
2 hi-lo	0.0537	2	0	0.0131	70	0	1.50
70 lo-hi	0.0130	7	0	0.0532	2	0	1.29
5 hi-lo	0.0196	13	0	0.0072	230	0	1.42
4 lo-hi	0.0074	230	0	0.0211	13	0	1.08
						average =	1.14

Source: Topper, Sandor, and Morrow, "Cumulative Fatigue Damage Under Cyclic Strain Control." Courtesy ASTM

should have had a value of 1.00, but the deviations are acceptable as being within normal scatter. It should be noted that the inherent stability of strain-controlled tests leads to better results than tests under stress control.

5.2 Plasticity and Cumulative Damage

It is useful to consider the accumulation of fatigue damage under any complex loading pattern by taking into account all of the cyclic plastic strains during the life [68]. To do this, let us consider Eq. 39 again, and derive from it a relationship for damage accumulation:

$$(\Delta\epsilon_p)(2N_f)^{1/(1+5n)} = \epsilon_f$$

where $-1/(1 + 5n) = c$, the slope of the $\Delta\epsilon_p$ vs. life plot. Since

ϵ_f is a constant,

$$(\Delta\epsilon_p)^{-1/c}(2N_f) = D, \text{ a damage number, a constant} \tag{41}$$

We can expect that $-1/c$ should vary between 1.4 and 2.0, because for most metals $-0.5 < c < -0.7$. The value of D is dependent on the monotonic fracture ductility, but it is worthwhile to consider it as an independent, constant quantity based on cyclic plastic strain, life, and the ductility exponent.

The simplified diagram in Fig. 76 helps in understanding the meaning of the independent quantity D. For a given material, c is a constant, and according to Eq. 41 the damage numbers D_A and D_B are equal. Note that in Eq. 41 the use of cycles or reversals, and of strain range or amplitude, is immaterial as long as it is consistent from one level of strain to the other.

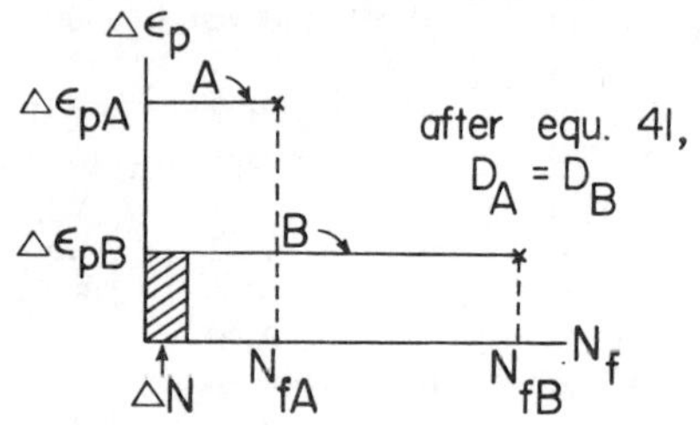

Fig. 76. Model of linear damage rates

The value of dealing with a damage number D is greatest in complex cumulative damage problems where the cyclic loading does not produce a constant plastic strain in each cycle. For background, consider the shaded area in Fig. 76, representing the damage done, or life exhausted, in ΔN cycles. Equation 41 gives the fractional damage number by raising $\Delta\epsilon_{p,B}$ to the $-1/c$ power and multiplying it by ΔN. Thus the calculated fraction of D is a measure of the life spent in terms of plastic strains. In multiple-level loading the prediction of life is possible without directly depending on extensive life data. The damage number D can be established on the basis of one known fatigue life, and damage summations can be made at each level of plastic strain. From Eq. 41, failure occurs when

$$\sum (\Delta\epsilon_{p,i})^{-1/c}(\Delta N_i) = D \tag{42}$$

where $\Delta\epsilon_{p,i}$ = plastic strain per cycle at the ith level, ΔN_i = number of cycles performed at the ith level, and D = result from

Eq. 41 using fatigue life at a single level of plastic strain.

To give an example of using Eq. 42, first consider the value of D for the data shown in Fig. 57:

$$2N_f = 100 \text{ at } \Delta\epsilon_p/2 \cong 4 \times 10^{-2}$$
$$c = -0.57$$
$$D = 100(4 \times 10^{-2})^{1.75} = 0.36$$

At another level

$$2N_f = 10^5 \text{ at } \Delta\epsilon_p/2 \cong 8 \times 10^{-4}$$
$$D = 10^5\ (8 \times 10^{-4})^{1.75} = 0.38$$

This is excellent agreement, and it is possible because the fatigue properties are firmly established through experiments. It should be noted that less accurately known material properties or less linear behavior than that above could easily lead to errors of a few hundred percent.

For an example of a step loading, assume that the same steel is used as above. If 20 cycles are imposed at $\Delta\epsilon_p/2 = 4 \times 10^{-2}$ the fraction of life exhausted is measured by

$$\Delta D_1 = \left(\frac{\Delta\epsilon_p}{2}\right)^{-1/c} (\Delta 2N_1) = (4 \times 10^{-2})^{1.75}\,(40) = 0.144$$

To find the remaining life, for example the number of cycles after which the metal will fail under the strains of $\Delta\epsilon_p/2 = 8 \times 10^{-4}$, a valid total damage number is needed. In this case it is reasonable to assume that $D = 0.37$. The remaining life is the difference of what was originally available and what was spent in the first block, $\Delta D_{\text{remaining}} = D - \Delta D_1 = 0.37 - 0.144 = 0.226$. Consequently, failure will occur after

$$\Delta 2N_2 = \frac{0.226}{\left(\dfrac{\Delta\epsilon_{p,2}}{2}\right)^{-1/c}} = \frac{0.226}{(8 \times 10^{-4})^{1.75}}$$

$$= 6 \times 10^4 \text{ reversals at the second level}$$

This is the same as the result of a linear damage accumulation analysis. Forty percent of the life was exhausted at the larger strains $(n_1/N_{f,1} = 40/100)$ which leaves 60 percent of the life at the smaller strains $(n_2/N_{f,2} = (6 \times 10^4)/10^5)$.

Equation 42 may be unnecessarily laborious for damage prediction when the loading is relatively simple, but it represents a valuable physical basis for dealing with complex loading patterns. Furthermore, in some cases the imposed loads may be constant, but the resulting plastic strains could vary because of severe softening or hardening or as a result of thermal treatments during the life [65]. In such cases the linear cumulative damage law based on stress vs. life or total strain vs. life curves could be entirely inadequate.

As an example of handling a complex problem, consider an imposed condition that causes a variation in plastic strains per cycle vs. progressive cycles, as in Fig. 77. The first problem is to establish

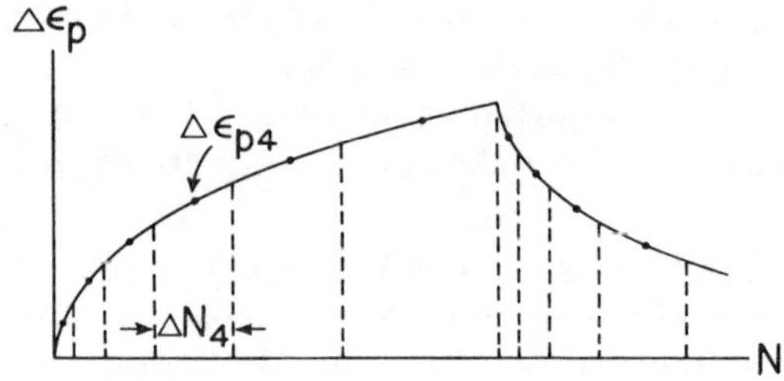

Fig. 77. Piecewise linear representation of damage under complex cyclic plastic strain

a valid damage number D according to Eq. 41, on the basis of a known constant c and a constant (or relatively so) $\Delta\epsilon_p$ for the given material. The summation of Eq. 42 can be produced either by cycle-to-cycle integration or by the approximation indicated in Fig. 77. In the latter case the cycles are grouped so that only small variations in $\Delta\epsilon_p$ occur within each group, and representative plastic strains are found for these groups. The summation becomes

$$(\Delta\epsilon_{p,1})^{-1/c}(\Delta N_1) + (\Delta\epsilon_{p,2})^{-1/c}(\Delta N_2) + (\Delta\epsilon_{p,3})^{-1/c}(\Delta N_3) + \ldots$$

Fracture will occur when the summation reaches the value of D representative of the material.

The use of a constant damage quantity such as that in Eq. 41 provides an opportunity to discuss the physical meaning of the so-called fatigue limit. The concept of the fatigue limit, or of the endurance limit, implies that there is a stress amplitude at which the S-N curve becomes horizontal so that fatigue failure does not occur in a finite life at stresses below this limit. At first sight any discussion of the reality of a fatigue limit appears to be entirely academic, because it is obvious that extremely long lives are involved even if the lives are always finite. There are two reasons for studying the problems associated with extremely long lives. First, there are practical situations, however rare, where very long cyclic and real-time lives must be considered. In some of these, for example with strain-aging metals, it is not realistic to obtain life data by accelerating laboratory tests. The other reason is that much can be learned about the overall fatigue problem by attempting to understand some exaggerated, hypothetical situations.

The bases for the present discussion have already been mentioned, and they are emphasized again. These are the facts that cyclic plastic strains are the cause of fatigue damage accumulation at all levels of loading, and that plastic strains result from the application of any finite stress in probably any material. The first of these two statements has been adequately supported by observed behaviors at short and intermediate lives, but not at long lives, where cyclic plastic strains have not yet been measured. There is evidence, however, that the concept of cyclic plasticity can be reasonably extended to the long-life region. Figure 78 shows that in several metallic and nonmetallic materials, the observable plastic strains depend only on the capability of the available instrumentation and not on the stress levels. It can be implied that the elastic limit (purely elastic behavior) is at zero stress [15].

An important aspect of the stress-strain data in Fig. 78 is the linearity of the log-log plots. This encourages extrapolation to strains other than those used in Fig. 78. In fact, most of the stress vs. plastic strain data available is above the range of strains shown. At those larger strains the behaviors are also linear.

The following sample problems involving large differences in cyclic plastic strains are based on the data for a mild steel [74], shown in Figs. 79 and 80. The damage number D is found from Fig. 80:

$$c = -0.6$$
$$\text{at } \Delta\epsilon_p = 10^{-1} \quad , \quad N_f = 55 \text{ cycles}$$
$$D = (\Delta\epsilon_p)^{-1/c}(N_f) = (0.1)^{1.66}(55) = 1.19$$

To check this number, select another point from Fig. 80; for instance, let $\Delta\epsilon_p = 10^{-2}$.

$$N_f = \frac{D}{(\Delta\epsilon_p)^{-1/c}} = \frac{1.19}{(10^{-2})^{1.66}} = 1.19 \times 2150 = 2560 \text{ cycles}$$

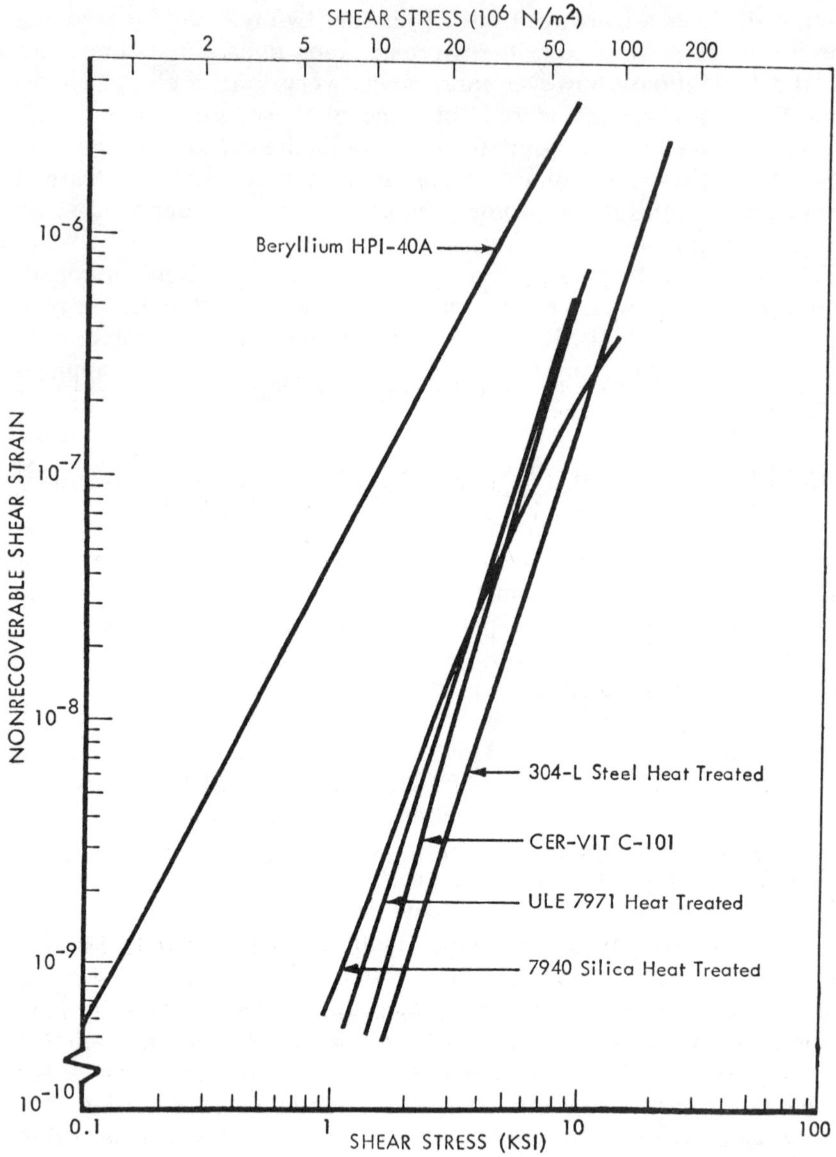

Fig. 78. Nonrecoverable strains at small stresses (from Eul and Woods, "Shear Strain Properties to 10^{-10} of Selected Optical Materials")

This is in good agreement with the life of 2400 cycles at the given strains as read from Fig. 80. Thus the damage number is valid at least for the short life region.

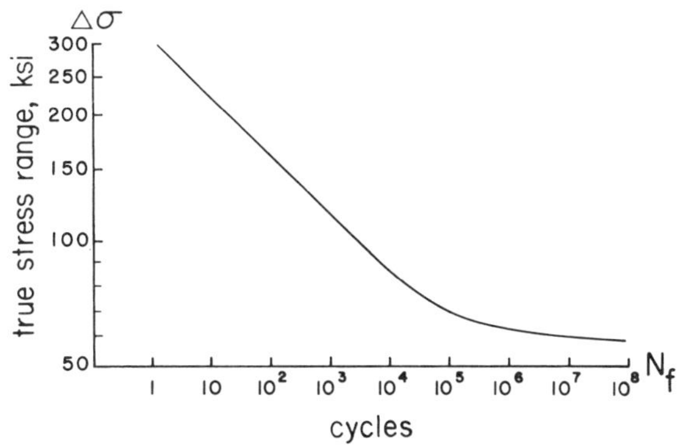

Fig. 79. Stress vs. life for mild steel (after Topper and Biggs, "The Cyclic Behavior of Mild Steel at Low Endurances")

In order to deal with the long life region, let us first extrapolate the straight line in Fig. 80 to any arbitrarily small plastic strain.

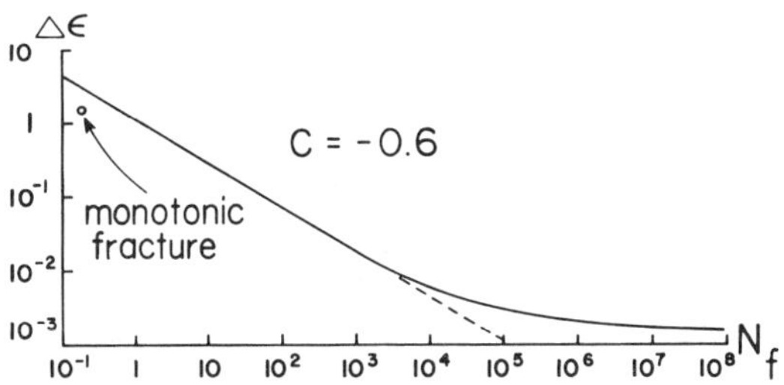

Fig. 80. Strain vs. life for mild steel (after Topper and Biggs, "The Cyclic Behavior of Mild Steel at Low Endurances")

For example, we get the life of 10^7 cycles at approximately $\Delta\epsilon_p = 7 \times 10^{-5}$. This life could also be obtained by using the damage number derived in the previous calculation. Other data in [74] show that such an extrapolation of the plastic strain vs. life relation is not unreasonable. It is found that at $N_f = 10^7$ the total strain range is 2×10^{-3} and the true stress range is 58 ksi (Fig. 79). The elastic modulus is 29.7×10^6 psi, so that the elastic strain range is

$$\Delta\epsilon_e = \frac{58{,}000}{29.7 \times 10^6} = 1.95 \times 10^{-3}$$

The plastic strain range is

$$\Delta\epsilon_p = \Delta\epsilon_t - \Delta\epsilon_e = 2 \times 10^{-3} - 1.95 \times 10^{-3} = 5 \times 10^{-5}$$

which agrees fairly well with the result of the extrapolation. Of course this calculation is not entirely reliable, because the particular stress and strain quantities are not known with sufficient accuracy. The implication is, however, that cyclic plastic strains could be used for damage analysis even where the *S-N* curve is not linear.

The simple examples can be carried to even smaller strains. The smallest cyclic plastic strains that can be conveniently measured are about 10^{-3} to 10^{-4}. At levels of one-thousandth or one-millionth of these values the fatigue lives become

$$N_f = \frac{D}{(\Delta\epsilon_p)^{-1/c}} = \frac{1.19}{(10^{-7})^{1.66}} \cong 10^{11.7} \text{ cycles at } \Delta\epsilon_p = 10^{-7}$$

and

$$N_f = \frac{1.19}{(10^{-10})^{1.66}} \cong 10^{16.6} \text{ cycles at } \Delta\epsilon_p = 10^{-10}$$

These estimates are tremendously large cyclic lives. They are finite, however, even though the cyclic stress amplitudes that may cause these lives must be well below the so-called fatigue limit of about 30 ksi in Fig. 79. Estimating the actual stresses is difficult for this mild steel, because the hardening mechanisms appear to be different for total strains above and below 0.05 [74].

A question could be raised, and rightly so, as to whether it is reasonable to assume a constant cyclic plastic strain throughout the life. In the above examples such an assumption was made by using the half-life plastic strains which are the most representative of the overall behavior. The results for the mild steel seem to justify the use of a representative, constant value for plastic strains, even though this metal is far from being cyclically stable. Nonsteady plastic strains will be discussed in the following, mainly to elaborate on the concepts of cyclic stress-strain behavior.

The natural processes of cycle-dependent adjustment are exponential functions of cycles. The rates of change from cycle to cycle depend on the magnitude of the imposed loading. Figure 81 shows that the higher the level of the imposed strains, the faster is the adjustment in the independently varying stress. At the smallest cyclic strains even a small change in cyclic stress is spread over a large number of cycles. It must be added that any cycle-dependent change

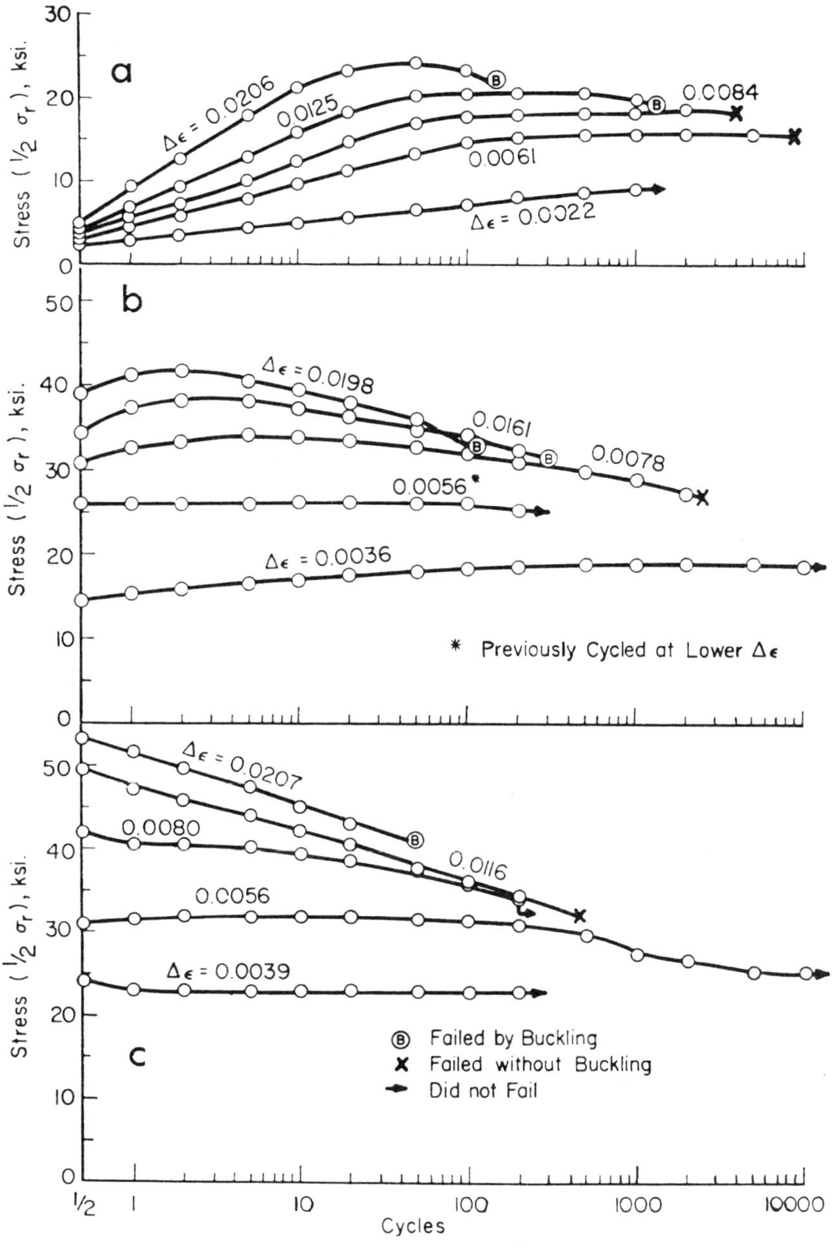

Fig. 81. Rates of change of cyclic stresses in copper: (a) fully annealed; (b) partially annealed; (c) cold-worked (after Tuler, in Morrow, *Internal Friction, Damping, and Cyclic Plasticity*. Courtesy ASTM)

can be expected only as a result of plastic flow and not of perfectly elastic, recoverable deformation. A formal statement of this behavior is that, for a given material, *self-induced changes in the cyclic stress-strain response tend to occur at rates directly proportional to the cyclic plastic strain at the particular point in life.*

In light of this statement let us re-examine the long-life behavior of mild steel. At a constant cyclic plastic strain of 10^{-7}, for example, the predicted life was $10^{11.7}$ cycles. It is entirely reasonable to assume that plastic strains are not constant over the whole life, but that they change exponentially with respect to cycles. The plastic strain is not zero during the first cycle, however, and it cycle-dependently changes with respect to this initial value. If the initial level of plastic strain is 10^{-7}, as in the example, the rate of subsequent change should be extremely low. One could say that the cyclic loads are so small that the material is nearly in its unloaded state, and one would not expect much to happen from cycle to cycle under such a condition.

The possibilities of exponential cycle-dependent adjustment are shown schematically in Fig. 82, where $\Delta\epsilon_{p,1}$ is the plastic strain

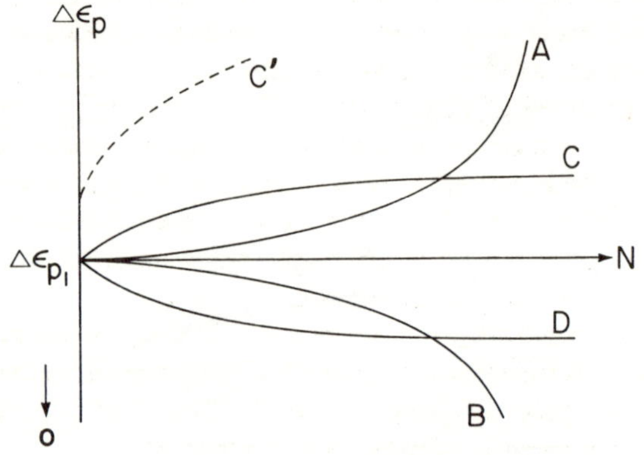

Fig. 82. Schematic functions of cycle-dependent plastic strains

in the first cycle. Curves *A* and *B* represent unstable material responses. Curve *A* is realistic for softening materials under stress control, and especially under load control, where runaway situations may develop. It is difficult to imagine, however, that curve *B* could exist for a real material. Such a behavior implies that the smaller the cyclic plastic strain becomes, the faster it drops toward zero. This contradicts the statement above that only a substantial cyclic plastic deformation can induce a high rate of change in stress-strain response.

Observations of material behavior lead to the belief that curves C and D in Fig. 82 represent very common responses, especially under strain control. At first thought, curve C may appear to be unreasonable because it is a cyclically softening pattern. In that case the increasing cyclic plastic strains may be expected to accelerate their own growth process. However, in any cyclic deformation there are competing processes of weakening and strengthening, and the observed behavior is always a net result of these two. One could say with respect to curve C that the material is weakening initially, but that there is a level of cyclic plastic strain which produces just enough strength by cold-working to stop further adjustments from cycle to cycle. This level is approached asymptotically. For such a behavior the spirit of the statement italicized above is satisfied even for different level control conditions. For example, the similar curves C' and C are for the same material, but the larger loads (C') cause the higher initial strain and the more rapid cycle-dependent adjustment.

Curve D in Fig. 82 represents cycle-dependent strengthening which can be caused by cold-working or strain-aging. The strengthening mechanism operates from cycle to cycle and requires prior plastic flow, but the hardening process is always a little faster than the weakening represented by plastic strain. Thus a stable level of cyclic plasticity is reached asymptotically. The position of the stable level and the rate at which it is approached again depend on the initial value $\Delta\epsilon_{p,1}$ which in turn is a measure of the imposed loading condition. At very low $\Delta\epsilon_{p,1}$ the stable level is reached only after many cycles, because in each cycle very little is happening (low $\Delta\epsilon_p$) that can lead to strengthening.

Fatigue life predictions can become very involved if accurate description of cyclic plastic strains as a function of cycles is required. In the general case of single-level loading but nonsteady plastic strains an expression is sought in the form of

$$\Delta\epsilon_p = Ke^p \tag{43}$$

where K = constant and p = function of N. The damage accumulation relation (Eq. 42) becomes

$$D = \int_0^{N_f} (Ke^p)^{-1/c}\, dN \tag{44}$$

The exponential function may have to be determined experimentally.

This could be very difficult when little plastic flow occurs. For example, the monotonic flow curve could give the initial value $\Delta\epsilon_{p,1}$ at the required stress level, but this alone is difficult for hard materials and for small stresses [15]. To find Eq. 43, cyclic tests are necessary at the desired stress amplitude. Using $\Delta\epsilon_{p,1}$ and $\Delta\epsilon_{p,N}$ after some N cycles, the needed constants can be evaluated for use in Eq. 44. The following examples involving exponential plastic strain variations are made with arbitrary constants.

5.3 Examples

Example 1

For a steel, $c = -0.6$, and at constant cyclic plastic strains of 0.05 the life is 200 cycles. These give a damage number $D = 1.365$. Find the life if the initial plastic strain is $\Delta\epsilon_{p,1} = 0.05$, but it changes cyclically so that

$$\Delta\epsilon_p = K\, e^{mN} \quad \text{where} \quad m = -\frac{1}{1000}$$

During the first cycle $\Delta\epsilon_p = 0.05$, and K can be evaluated at $N = 1$,

$$0.05 = Ke^{-1/1000} \quad , \quad K = 0.05\, e^{-1/1000}$$

According to Eq. 44

$$D = K^{-1/c} \int_0^{N_f} e^{-mN/c}\, dN = -\frac{K^{-1/c}c}{m}\left[e^{-mN_f/c} - 1\right]$$

The life can be expressed as

$$N_f = -\frac{c}{m} \ln\left(1 - \frac{Dm}{cK^{-1/c}}\right)$$

$$N_f = -\frac{(-0.6)}{(-1/1000)} \ln\left[1 - \frac{(1.365)\,(-1/1000)}{(-0.6)\,(0.00695)}\right] = 238 \text{ cycles}$$

This is qualitatively reasonable because the cyclic plastic strain is a decreasing function of cycles but at a relatively slow rate, so that the life should be slightly longer than 200 cycles obtained at steady strains.

Example 2

The above problem with an increasing exponential should give a reduced life. With m = 1/1000 and everything else unchanged, the life becomes 170 cycles. This is again reasonable, considering the slow rate of increase of $\Delta\epsilon_p$ for N $\ll$ 1000.

Example 3

A similar problem with $m = 1/1000$ and $\Delta\epsilon_{p,1} = 10^{-7}$ gives a surprisingly short life:

$$N_f = -\frac{(-0.6)}{(1/1000)} \ln\left[1 - \frac{(1.365)(1/1000)}{(-0.6)(10^{-11.65})}\right] = 12{,}500 \text{ cycles}$$

However, this is caused by the arbitrarily selected plastic strain function rising rapidly for $N > 1000$. In fact, it is incorrect to use the same function for different levels of loading as was done above.

Example 4

Another kind of difficulty arises when the last example is changed slightly. For $m = -1/1000$ and $\Delta\epsilon_{p,1} = 10^{-7}$ the life cannot be evaluated because the quantity $[1 - (Dm/cK^{-1/c})]$ becomes negative. This situation is the same as that shown by curve B in Fig. 82, where the plastic strains are decreasing at an accelerating rate. In such a case, failure will occur either early in the cycling or not at all. The improbability of such material responses has been discussed previously.

There are two lessons to be learned from the foregoing. In most cases it should be adequate, and certainly simplest, to use a constant, representative value for cyclic plastic strains in damage calculations. Whenever a more complex plastic strain function is reasonable, extreme care must be exercised in establishing the validity of the function before the damage summation is performed.

The conclusion to be drawn from this chapter is that a physical basis of fatigue damage accumulation must be, and can be, found for cumulative damage analysis. This basis is most likely to involve cyclic plastic strains, *and the method that emerges is a unifying one for all ranges of fatigue lives.* It must be emphasized that extra caution is in order when the fatigue problems include such complicating factors as mean stresses, mean strains, complex geometries, and unusual environmental conditions. To gain further understanding of the problems of fatigue damage accumulation consider [8, 20, 24, 79].

6

Mean Stresses and Mean Strains

In the preceding discussion of damage accumulation and damage rates, it was always assumed that the cyclic stresses and strains were fully reversed with respect to the virgin stress-strain state of the material. Thus according to the earlier discussed Figs. 23 and 24, σ_m and ϵ_m were always zero, regardless of how the imposed loading limits or the material's response to these conditions may have changed during the life. It is important to realize that real fatigue problems are seldom that simple and that some mean stress or mean strain or both is normally present. Another important fact is that mean stresses or strains are just as likely to change in a cycle-dependent fashion as the stress or strain amplitudes. All of these changes are caused by cyclic plasticity.

One way to look at the mean values is to consider them as static values upon which fully reversed alternating conditions are superimposed. In this respect the already mentioned quasi-static properties (Figs. 14–17) are most useful. The following examples of situations that can develop show the basic role of plastic deformations in creating *or* eliminating mean stresses and strains.

Figure 83*a* is a stress-strain diagram for the ideal, entirely elastic behavior. The material is preloaded to point *A* and cycled between the upper limit *B* and the lower limit *C*. The mean stress σ_m and the mean strain ϵ_m are the coordinates of point *A*, which represents the preload. If there is no cyclic plastic action, the quantities in Fig. 83*a* remain unchanged by cycling indefinitely. The situation would be similar in the third quadrant under fully elastic behavior.

With plastic deformations the behavior becomes more complex. In Fig. 83*b*, for the first example, tensile mean strain and zero mean stress is generated. The material is strained to *A*, unloaded to the mean strain at *B*, and cycled between *C* and *D*. Figure 83*c* is the way to produce zero mean strain and a compressive mean stress at *C*. The cycling limits are *B* and *D;* in this diagram they are chosen as stress control limits, but they could also be strain limits as shown in the next drawing. In part *d*, one way of generating a compressive mean strain and a zero mean stress is shown. The material is loaded to *A*, unloaded to the mean strain at *B*, and cycled between the chosen strains at *C* and *D*. The sketch in Fig. 83*e* demonstrates the production of a tensile mean stress at *C* without a mean strain. A similar pattern in part *f* shows that the two cycling limits are not necessarily of the same sign when a non-zero mean condition exists. It can be seen, however,

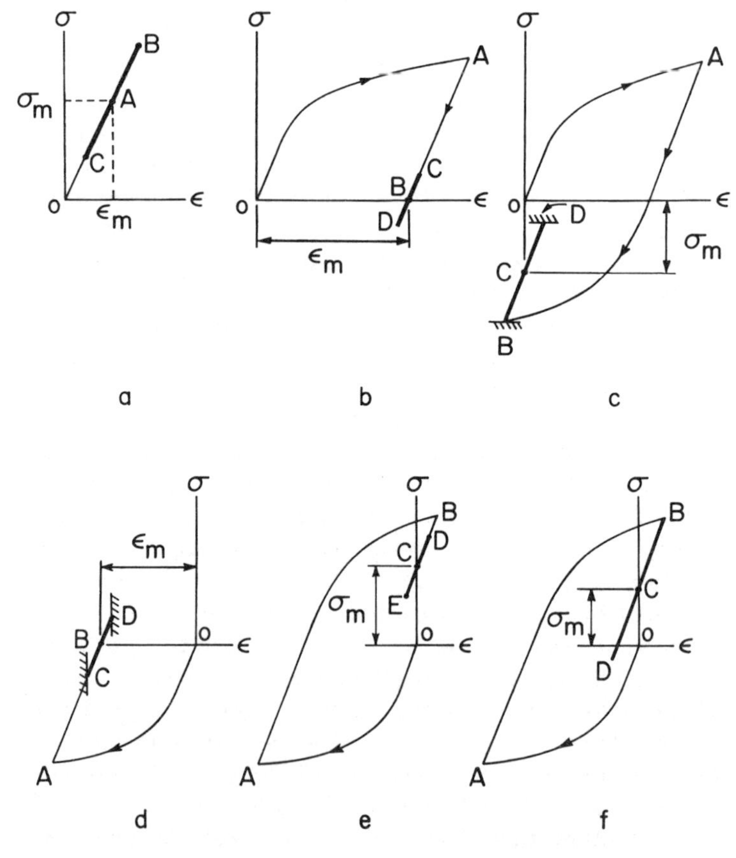

Fig. 83. The generation of mean stresses and mean strains

that for a given material the condition in part *e* could lead to a higher mean stress than that in part *f*.

The various diagrams in Fig. 83 should make it clear that it is actually difficult to return the material to zero stress and zero strain after some plastic flow has occurred. In fact, the only way to do this requires that the stress-strain path intersect the elastic line that goes through the origin of the stress-strain axes. For cycling with a closed hysteresis loop this means that the origin must be within the loop. Figure 84 shows how the original stress-strain condition can be reached from a large loop. If the diagram is a recording of the stress signal as a function of the strain signal, the loop is continuously written in the clockwise direction. To return to the origin, the cyclic loading must be interrupted at either point

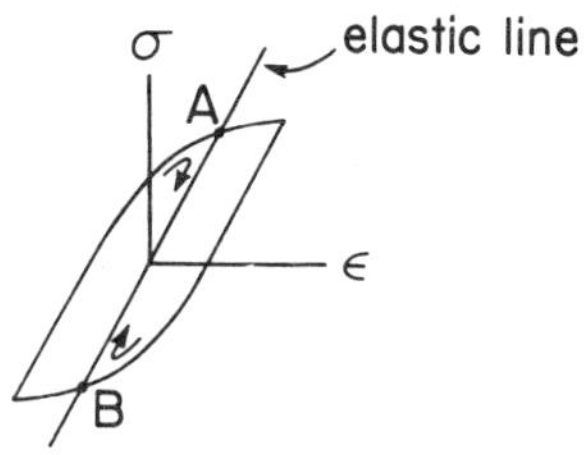

Fig. 84. Sole ways of returning to zero stress and strain after plastic deformation.

A or point *B*. A complete unloading from either of these two points produces the desired result.

The role of plastic strain in creating various magnitudes of mean strain can be seen in parts *b* and *d* of Fig. 83. The possible dependence of the magnitude of a mean stress on the amount of prior plastic strain is less obvious. Figure 85 shows various compressive mean stresses that can develop (here without mean strains) upon prior overstraining in tension, for example. A loading to point *A* can lead only to a small mean stress, which in this case cannot exceed the magnitude $\overline{OB}$. Of course, continuing the compressive loading from *B* along the dashed line can result in larger mean stresses, but this will be accompanied by large mean strains as well. Unloading from point *C*, on the other hand, can lead to point *D*, which would represent the mean stress in the limiting case of negligible cyclic stress range. Stopping at point *E* or *F* allows larger stress ranges at lessening but still considerable mean stresses. It is generally true that large stress amplitudes and large mean stresses are not compatible with each other.

The stress-strain diagrams with mean stresses that have been discussed so far do not show cyclic plastic strains after the mean

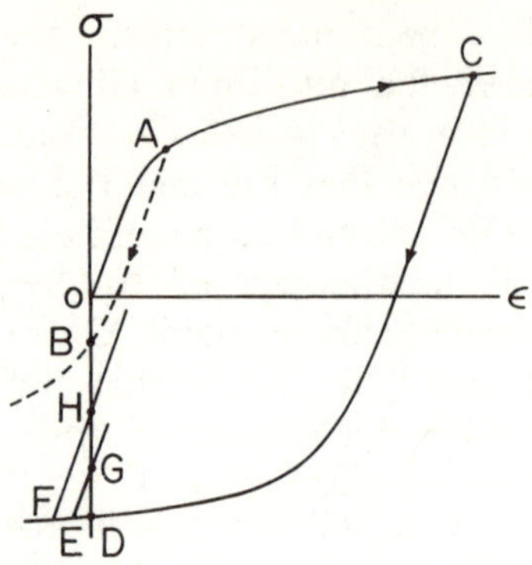

Fig. 85. The dependence of mean stress on prior plastic strain and on applied stress range; for D, $\Delta\sigma \approx 0$, σ_m = max.; for G, $\Delta\sigma > 0$, $\sigma_m <$ max.; for H, $\Delta\sigma \gg 0$, $\sigma_m \ll$ max.

values have been established. The hysteresis loops are present in these drawings, but they are pencil-line thin, indicating mainly elastic action. This is not always the case, since it is possible to have a mean stress and sizable cyclic plastic strain in the same material region at the same time. It is important to realize, however, that it is not possible to make both of these quantities simultaneously large in any material. The reasons for this can be understood from the following.

A simple explanation is provided by Fig. 86. Assume that a material is about equally strong in tension and compression, so that stresses exceeding the levels A or B result in failure. It is clear that the largest possible cyclic plastic strain is obtained when the largest possible stress range is imposed on the material. This is equivalent to saying that the stress limits should be just a little less than σ_A and σ_B for maximum plasticity. If we want to cycle about

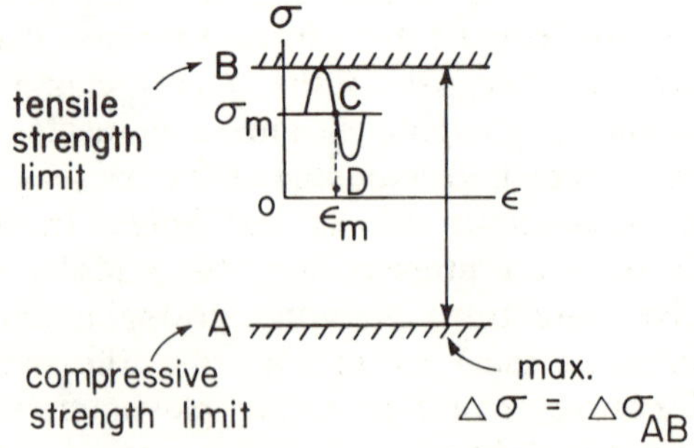

Fig. 86. The incompatibility of large mean stress and large cyclic stress; note that the sine wave is not correct in the σ-ϵ plane

point C, which has a high mean stress, the proximity of C to level B will determine the maximum allowable stress amplitude. Since a mean value is always a halfway point between the peaks, the maximum stress range that the material will tolerate is shown by the sine wave in Fig. 86, and its magnitude is $2(\overline{BC})$. Obviously this is a small cyclic load compared to $\Delta\sigma_{AB}$ and the expected plastic strains for cycling about point C are negligible. Using a similar argument, it could be shown that for cycling about point D the maximum possible cyclic plastic strains are relatively large because the mean stress is close to zero.

Another way to look at the concurrent mean stress and cyclic plastic strain quantities is to rely on the fundamental concept of cyclic plasticity. By definition, this means reversed plastic flow, which is never produced by merely unloading the material. Figure 87 is an example of a situation where cyclic plasticity is not possible, because the stresses are always of the same sign so that the plastic flows (if there are any) cannot be reversed. In this case, the only cyclic load that the material experiences is a tensile loading from C to B, followed by unloading from B to C.

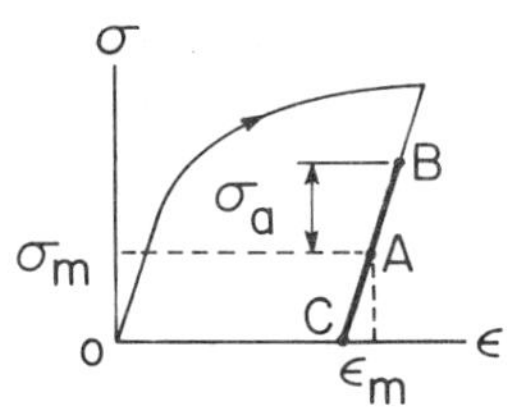

Fig. 87. Elastic behavior with mean stresses and mean strains

It might be asked whether, if cyclic plasic strains are not possible, the condition shown in Fig. 86 (σ_m at C) could ever lead to a fatigue failure. The answer is that failure is entirely possible, and in principle it is still caused by plastic strains. These strains are not obvious, however, because of the mostly elastic behavior. The plastic deformations are highly localized under small loads, and occur mainly in the neighborhoods of stress raisers, such as inclusions and even microscopic cracks and holes. In these small but critical neighborhoods the stresses may be partially in the reversed (here compressive) direction, even though macroscopically the imposed condition is as ideal as in Fig. 87. The partially reversed stresses do, of course, result in cyclic plastic flow, and eventual failure becomes a possibility.

The generalized conclusion from these considerations is that, under mean stresses, cyclic plastic strains will be present only if the stress limits alternate over the tensile and compressive halves of the stress plane. The half planes of stress are mentioned here because, basically, it does not matter where on the strain axis these stress alternations occur. The hysteresis loop representation of these ideas is shown by four examples in Fig. 88. Similar observations can be made under random loading as seen in Fig. 16 of [12].

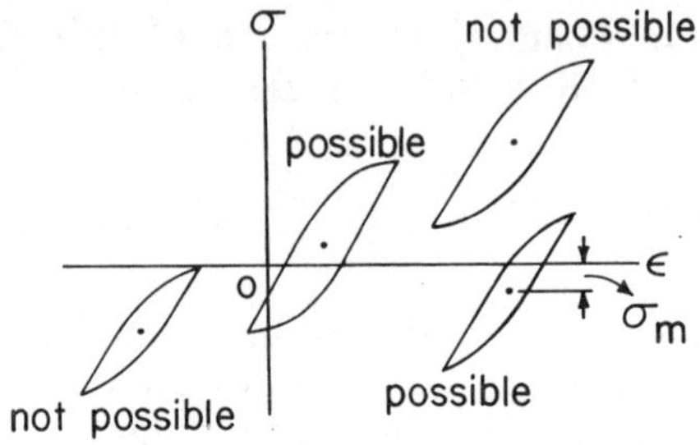

Fig. 88. The incompatibility of large mean stress and large cyclic plastic strain

6.1 Fatigue Life with Mean Strains

A mean strain or prestrain, as in Fig. 83*b*, can under certain conditions reduce the subsequent fatigue life by exhausting some of the ductility available in the virgin material [7, 40, 75, 81]. An important aspect of mean strains is related to mean stresses, but the combined problem is easier to understand on the basis of separately discussed phenomena. Assume, for the time being, that mean stresses have nothing to do with mean strains.

Mean strains can contribute quite directly to the damage that leads to failure, as shown schematically in Fig. 89. The monotonic fracture ductility ϵ_f is assumed to be equal to the fatigue ductility ϵ_f'. A prestrain to point *A* exhausts a large fraction of the available ductility. For cycling about *A* let us imagine that there is no cyclic plasticity but that somehow this cycling causes point *A* to shift

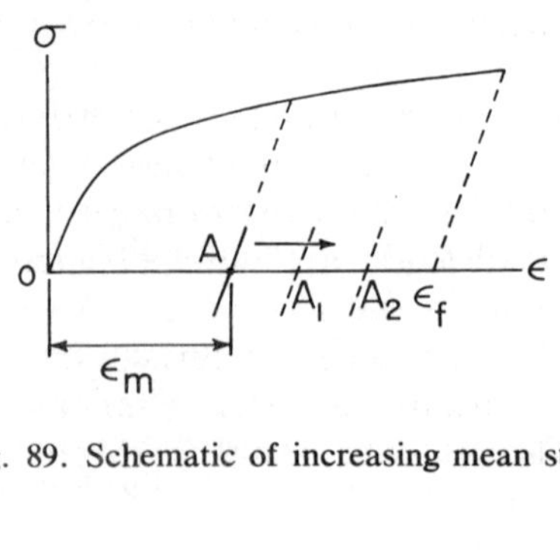

Fig. 89. Schematic of increasing mean strain

toward the right. As the mean strain increases (through $A_1, A_2, A_3, \ldots$), more and more of the ductility is exhausted, and failure occurs when ϵ_m equals ϵ_f. If we also consider the presence of cyclic plastic strains as A shifts to the right, then failure may occur before ϵ_m reaches ϵ_f. Whether the mean strain is stationary throughout the life or not, it can contribute to the processes that exhaust the ductility and consequently the remaining life. A formal but idealized expression of this fact can be made by modifying Eq. 39 to include a mean strain:

$$(\Delta\epsilon_p)\,(2N_f)^{1/(1+5n)} = \epsilon_f - |\epsilon_m| \tag{45}$$

where the absolute value of the mean strain (or a cyclic prestrain) is used because both tensile and compressive deformations can lower the ductility. There are indications that in low-cycle fatigue, the effects of tensile and compressive mean strains are quite similar to each other [54].

For the prediction of fatigue life, the proposed modification appears to be a simple one that leads to a minor change in the damage value D in Eq. 41. In reality, the matter is more complex than that, as will be seen in the following.

It has been observed [41, 75] that inelastic prestrains may cause a relatively large reduction in the subsequent fatigue life at small loads. This can be more severe than the reduction expected on the basis of partial damage directly attributed to the prestrains. A behavior like that illustrated in Fig. 90 is rather surprising, even though here the prestrains are cyclic without leaving a mean strain. The generalization can be made that the large reduction in life at small loads can be caused by either cyclic or monotonic prestrains. The important point is that the magnitude of the reduction cannot always be accounted for by the linear damage summation procedures. From the practical point of view, it would seem reasonable to apply a small prestrain to all specimens before testing for design data.

There are several possible reasons for the high rate of damage after prestrains. One plausible explanation [41] is that the large prestrain initiates cracks that can propagate at the smaller loads. This implies that small loads are more efficient in making a crack grow than starting one in the first place. Another explanation (not applicable to Fig. 90) concerns materials that show sudden yield phenomena. Plastic strains in these eliminate the yield point at once and cause softening. The result is accelerated damage through

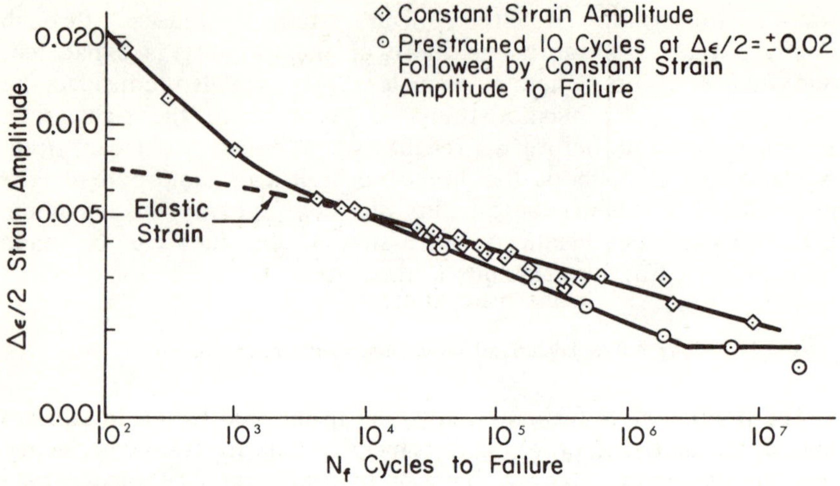

Fig. 90. The effect of prestrain on fatigue life (from Topper and Sandor, in *Effects of Environment and Complex Load History on Fatigue Life*)

enhanced cyclic plasticity under any common control condition. Still another explanation will be given after the discussion of mean stress relaxation.

The exact amount of prior plastic straining that reduces the life by a certain factor at nominally elastic strain levels is not well known. A moderate amount may be generally adequate to cause a noticeable change [75]. As a rule of thumb, it may be assumed that a plastic prestrain of about 0.01 (1 percent) or more is significant in the fatigue life of metals.

6.2 Fatigue Life with Mean Stresses

Some of the earliest observations of the fatigue phenomenon showed that mean stresses can significantly alter the resistance to cyclic loads [22]. Nevertheless, the largest part of the published fatigue data was obtained from fully reversed cycling. This was a matter of test equipment: the most widely used rotating bending machines did not allow the application of mean stresses.

The behavior in the presence of mean stresses is such that tensile mean stresses reduce and compressive means increase the life at a given amplitude of loading. On an *S-N* diagram a family of curves can be plotted with each line corresponding to a different mean stress, as shown in Fig. 91. A similar pattern is found in strain vs. life plots, as in Fig. 92. The figure also indicates that the curves are not necessarily parallel to each other.

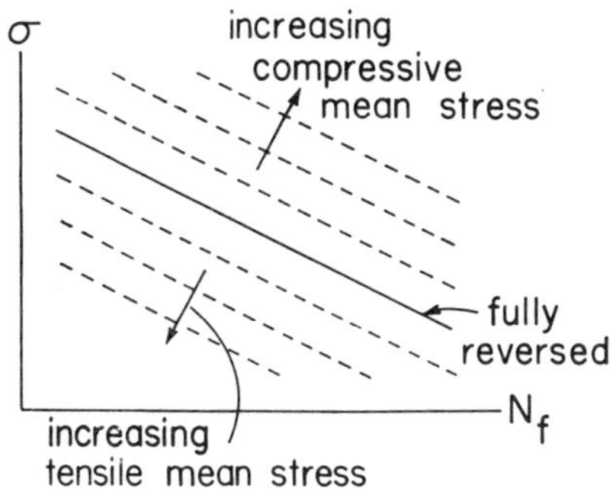

Fig. 91. Schematic of fatigue lives with mean stresses

The prediction of fatigue strength as influenced by mean stresses cannot yet be based on physical principles that are fully convincing. Several empirical relations have been proposed [19], and these are adequate for the handling of relatively simple problems. All of these relations are based on the concept of superimposing an alternating stress on a static stress and attempting to find limiting values in fatigue for the combination of these stresses. The basic argument is similar to the one used previously in discussing the incompatibility of large mean stresses and large cyclic plastic strains.

An idealized description of the superposition of stresses can be given with the aid of Fig. 93. The limiting value of any combination of stresses is σ_f, the true fracture strength in monotonic tension.

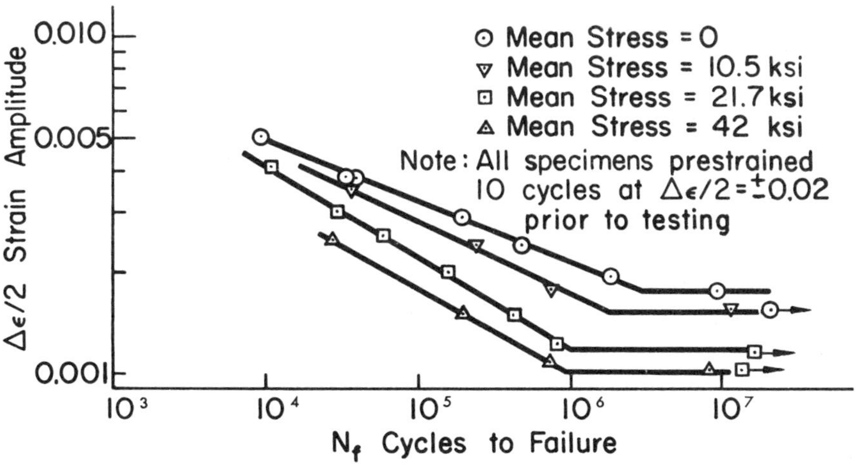

Fig. 92. Examples of fatigue lives under tensile mean stresses (from Topper and Sandor, in *Effects of Environment and Complex Load History on Fatigue Life*)

Ultimate strengths, σ_u, and yield strengths, σ_y, are also possible and often-used arbitrary limitations, but σ_f is the largest allowable stress. According to Fig. 93, a zero mean stress allows a maximum alternating stress that approaches σ_f. Of course, when $\sigma_a = \sigma_f$ we should not expect a life of more than one quarter of a cycle. The reverse is also true, so that at mean stresses approaching σ_f the allowable alternating amplitude superimposed on the mean stress is nearly zero. For an ideal behavior we could expect that the relation $\sigma_a + \sigma_m \leq \sigma_f$ must always be true, resulting in a straight line between points *A* and *B* in Fig. 93. For example, a mean stress of $\sigma_f/3$ would determine the maximum allowable alternating stress as $\frac{2}{3}\sigma_f$.

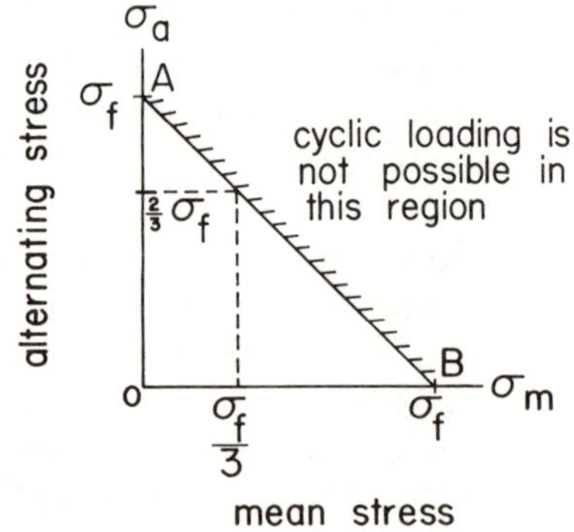

Fig. 93. The idealized effect of mean stresses

Such idealized behavior could be valid only if damage did not accumulate from cycle to cycle. In this ideal case the fatigue problem would not exist at all. In reality, the cumulative nature of the damage makes the approach based on true fracture strength too liberal. That is unfortunate because it forces us to abandon a simple approach to the mean stress problem in fatigue. At the same time, the search for a new physical basis to describe the real behavior runs into extreme difficulties. Numerous mathematical expressions have been proposed. None of these is perfect, but they are much better than ignoring the mean stress.

The problem has been approached in a practical way by developing empirical relationships. In essence, these are similar to the concept presented in Fig. 93. The three most widely used relations [19] are shown in Fig. 94. Such plots are often called Goodman diagrams. The ordinate of the diagram is the fully reversed alternating stress

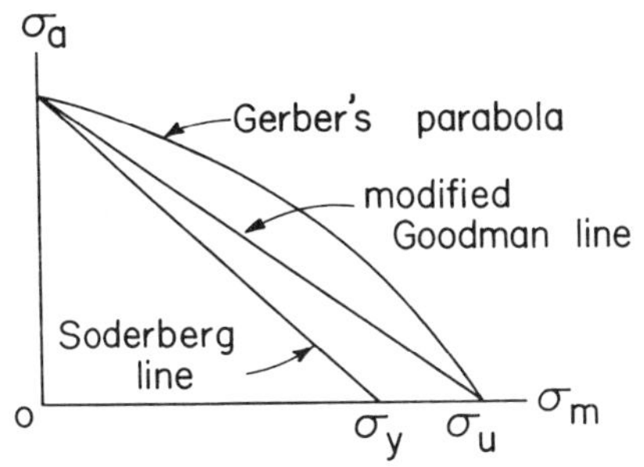

Fig. 94. Three approaches to design with mean stresses (after Forrest, *Fatigue of Metals*)

σ_a for a chosen life endurance. The limiting maximum mean stress is arbitrarily chosen as either the ultimate strength σ_u or the yield strength σ_y. Note that both of these are below the true fracture strength. Lines connecting the selected points on the two axes represent combinations of mean and superimposed alternating stresses that give the same endurance as the selected fully reversed amplitude. The curves may be determined experimentally after establishing a number of *S-N* curves, each for a constant value of σ_a, σ_m, or the stress ratio $R = \sigma_{min}/\sigma_{max}$.

Considering the fact that a material may cycle-dependently change its stress-strain response, the empirical relations are quite arbitrary and sometimes inadequate. For example, the Soderberg line in Fig. 94 is a statement of the condition that yielding should not be allowed for any combination of mean and alternating stresses. There is a definite danger in attempting to use the proposed relationship without proper consideration for the material's cyclic behavior. A quick comparison of the particular monotonic and cyclic stress-strain curves should indicate whether the use of monotonic yield strength is reasonable or not. In general, even empirical relations of mean stress effects should be reexamined in the light of the fact that large changes in deformation resistance are possible in fatigue.

The simplest way to account for cycle-dependent material properties and mean stresses [35, 49] is similar to the concept of the maximum allowable stress discussed with respect to Fig. 93. We could say that the alternating stress plus the mean stress could never exceed the true cyclic fracture strength. In a simplified manner, this can be expressed formally as a change in the fatigue strength coefficient that is caused by a mean stress. Thus on the basis of Eq. 38,

$$2N_f = \left(\frac{\sigma_a}{\sigma_f' - \sigma_m} \right)^{1/b} \tag{46}$$

Since b is negative, a tensile mean (+) reduces the life, and a compressive mean (−) increases it. In some cases Eq. 46 can be used to account for the effect of mean stresses, as shown in Fig. 95. This is not always true, however, and care must be exercised when using the equation.

It may appear at first sight that Eq. 46 represents the desired physical basis for dealing with mean stresses in fatigue. Such a thought may be strengthened by data like that in Fig. 95 and by the fact that Eq. 46 is easily related to the limitation imposed by the monotonic fracture strength (σ_f in Fig. 93). In reality, the situation is complicated somewhat by several factors. First of all, the true fracture strength is not a constant that is entirely independent of length of life. Equation 46 is based on such constancy, but according to considerations of fracture mechanics, the fracture strength depends on the fatigue crack's size (Griffith relation, Eq. 20), which in turn depends on the life range. Other practical problems may result from slight changes in the exponent b and from the effect of cycling on the mean stress itself. The significance of some of these depends on the magnitudes of the static and alternating stresses and on the specimen size. For instance, one could expect a reasonable prediction by using Eq. 46, if σ_m is small relative

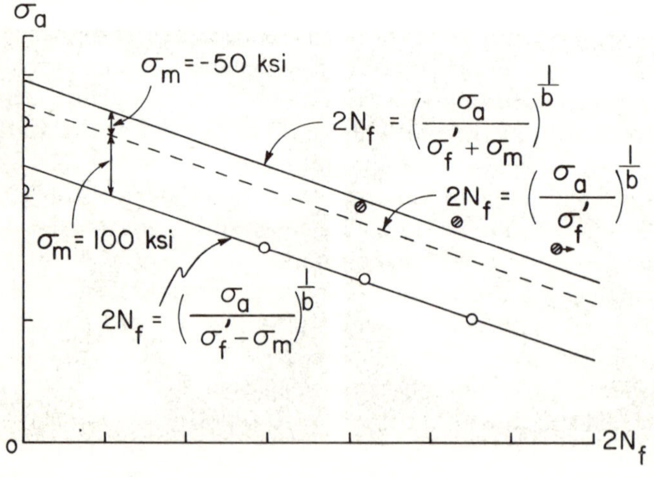

Fig. 95. A simple analytical method of accounting for mean stresses (after Landgraf, "Effect of Mean Stress on the Fatigue Behavior of a Hard Steel")

to σ_f' and if the specimen is so small that the fatigue cracks developing at various lives could not be too different in size from each other.

There is some evidence that the size of usual laboratory fatigue specimens will not lead to significant errors caused by different crack sizes. Figure 96 shows the fracture surfaces of 0.25 inch diameter axial fatigue specimens tested at two stress levels. The lower stress allows the development of the larger fatigue cracks, but the stress at fracture appears to be the same in both specimens. The approximate fracture stresses are calculated by dividing the actual load at fracture by the final load-bearing area. The latter is difficult to measure accurately, hence the approximation. In precise fatigue analysis of large structures the true fracture stress should be taken into account, because the size of the members could allow crack growth to large dimensions.

6.3 Fatigue with Mean Stress and Mean Strain

The simultaneous existence of both mean quantities is a common occurrence. Strictly speaking, it is not possible to hold both values at zero. It may be argued that approximately zero means are adequate in practice when required, but it will be shown that even negligible mean values could lead to undesirable developments.

The combined effects of the mean values are basically caused by mean stresses, especially where considerable cyclic plasticity is occurring. A tensile mean stress could cause the mean strain

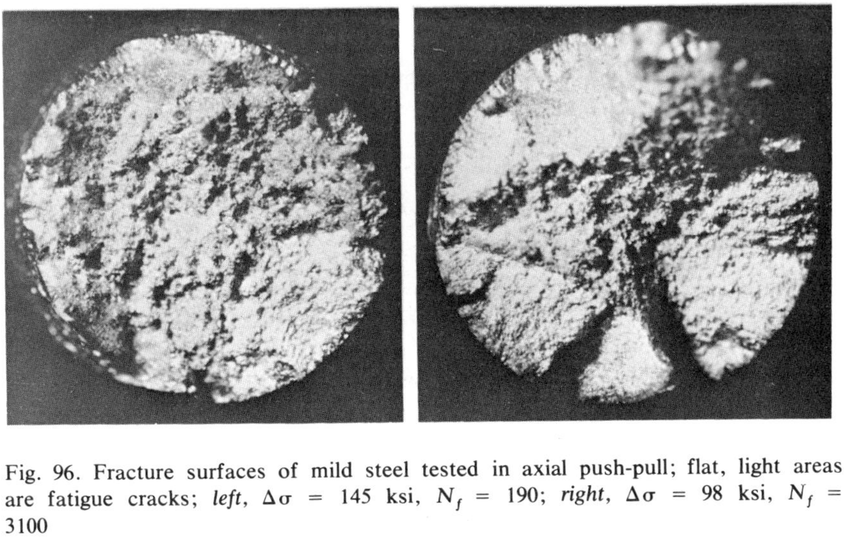

Fig. 96. Fracture surfaces of mild steel tested in axial push-pull; flat, light areas are fatigue cracks; *left*, $\Delta\sigma$ = 145 ksi, N_f = 190; *right*, $\Delta\sigma$ = 98 ksi, N_f = 3100

to gradually increase in the tensile direction. On a stress-strain diagram it would appear that the hysteresis loop is "walking" or "cyclically creeping" in the direction of increasing tensile strains. Such changes are dangerous, because the mean strain could approach the fracture ductility of the material and lead to failure by excessive tensile strain. The most undesirable situations occur with cyclically softening materials under load control, and under either tensile or compressive mean stresses. Figure 97 shows the problem with a tensile mean stress. Here the damage accumulation is being accelerated in two ways. First, the cyclic plastic strain increases from cycle to cycle because of the softening. Second, there is cyclic creep which pushes the mean strain toward the maximum available ductility.

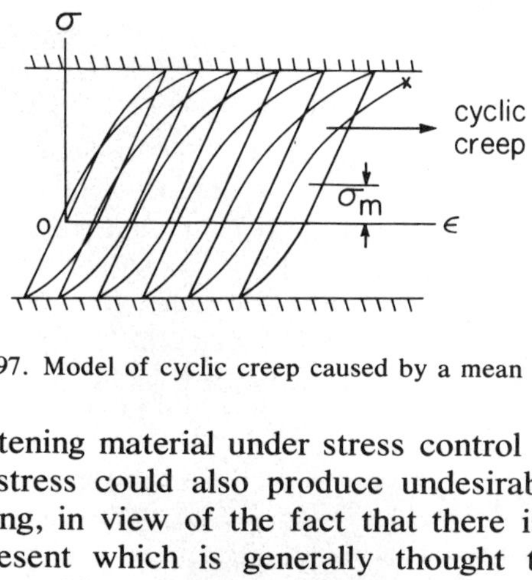

Fig. 97. Model of cyclic creep caused by a mean stress

The same softening material under stress control but with a compressive mean stress could also produce undesirable results. This may be surprising, in view of the fact that there is a compressive mean stress present which is generally thought to be beneficial in fatigue. The problem is in the cyclic creep, which is now in the compressive direction. The cycle-dependent softening gradually reduces the tangent modulus and the compressive mean stress enhances the conditions for buckling. Figure 98 is the stress-strain

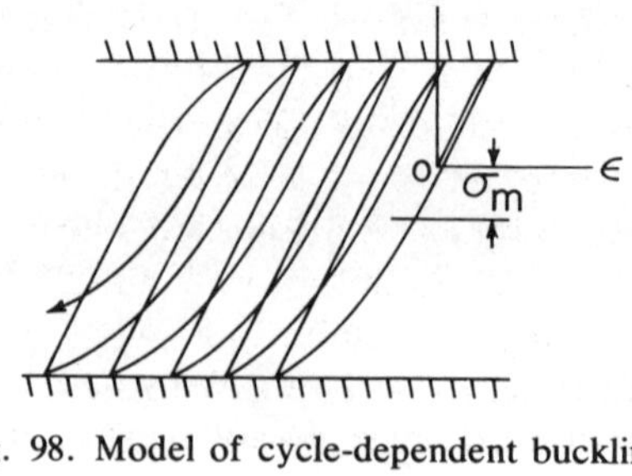

Fig. 98. Model of cycle-dependent buckling

representation of the events leading to cycle-dependent buckling. Of course, this diagram does not show the local conditions that exist when bending finally occurs under the originally axial loads. The important point is that a statically adequate member may gradually become critical under cyclic loading, as a result of either tensile or compressive mean stresses.

The results can be quite interesting when the conditions described above are applied to a cyclically hardening material. In such a case, if the loads are high enough there is cyclic creep, but its rate is decelerating. Figure 99 shows that a reduction in cyclic plastic strain is accompanied by a stabilization in the tensile mean strain. Similar stabilization, for this metal, can be expected from compressive mean strains under compressive mean stresses.

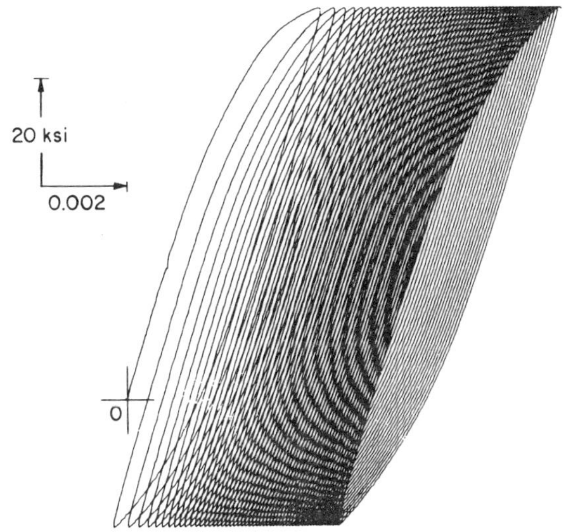

Fig. 99. Deceleration of cyclic creep (from Landgraf, in *High Fatigue Resistance in Metals and Alloys*. Courtesy ASTM)

The possible modes of failure for a hardening material are the same as for a softening one. Tensile fracture can occur by cyclically induced creep before stabilization is attained, and progressive buckling can occur in compression. For a cyclically hardening material, however, the matter may become a race between the harmful and beneficial processes. For example, the combination of load range and hardening rate may be such that even under a tensile mean stress, the mean strain can stabilize before much cyclic creep occurs. Similarly, under a compressive mean stress, the combination may produce a sufficiently high tangent modulus and may stop the compressive creep before the material has experienced a significant amount of nonaxial deformation that could enhance buckling.

It must be noted that cyclic creep can be caused not only by mean stresses but even under fully reversed loads, if the material is not equally strong in tension and compression [34, 62]. One may say, in that case, that the material applies its own effective mean stress. Examples of such materials include cast iron, maraging steel, and composite materials in general.

The rate of cyclic creep is intimately related to the amount of cyclic plasticity. The reason for cyclic creep is that in each cycle the mean stress raises the peak stress in one direction only. Thus in each cycle there is some plastic flow that is not opposed by an equal amount of flow in the other direction. With a high load range, the cyclic plastic strain may be high even without any mean stresses. In this case, a very small mean stress can produce substantial preferential plastic flow in each cycle, and cyclic creep results [6].

As an example of the behavior described above, consider an axial loading of mild steel. At cyclic stresses of ±70 ksi the life is about 200 cycles. With an initial tensile mean stress of only 1 ksi, a runaway situation with obvious cyclic creep may develop, cutting the life in half. For another example, see Fig. 9 in [13]. It must be noted that the initially imposed mean stress would not be constant throughout the life without constant adjustment. This is true because with the cyclic creep in tension, the specimen's cross-sectional area gradually decreases. This makes the stresses increase above the nominal or initial values, if instead of true stress control there is cyclic-load control.

With very small cyclic plastic strains the rate of cyclic creep should be extremely small. A factor that could offset this advantage is that large mean stresses are possible under small cyclic loads. In any case, it is still true that all cycle-dependent changes occur over many cycles when the plastic strains are small.

6.4 Theory of Fatigue Damage with Mean Values

Ideas have already been presented concerning the contributions of prestrains to the fatigue-damage process. Mean stresses can be appreciated best in connection with stress concentrations. Consider, for instance, the small fatigue crack or other flaw in Fig. 100. Under a perpendicular tensile-stress field this crack tends to open, as in part *b*. A compressive stress tends to close the crack and

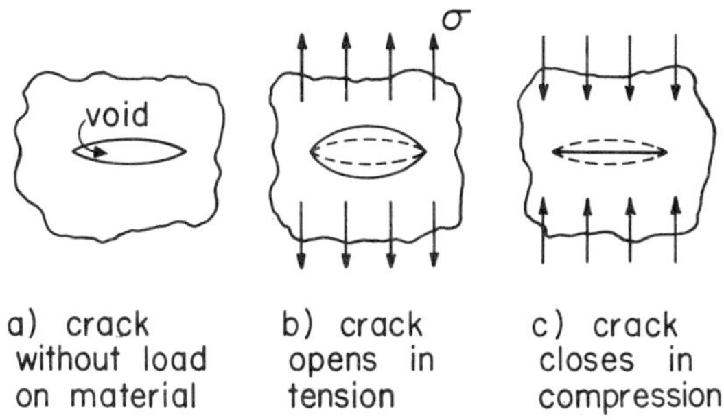

Fig. 100. Model of changing crack shape in fatigue

may even bring the faces into contact with each other, as in part *c*. The sharp tips of the crack are points of severe stress concentration as long as the crack is open. Under compression, the stress concentration tends to be reduced, or sometimes even eliminated effectively as the crack closes. In fatigue, the crack is cycled between relatively open and relatively closed positions. When a tensile mean stress is superimposed on the alternating stress, the crack is in the opening mode most or all of the time. This way the stress concentration is generally active, whereas under compression it would be generally inactive.

Because of active stress concentrations, every small flaw in the material is surrounded by a region where the cyclic stresses and strains are larger than the average (nominal) values for the given conditions. An extreme example of this would be a situation in which considerable plastic strains occur at the tips of cracks, while most of the material is stressed well below gross yielding. In less extreme circumstances (highly unlikely), all regions may be far from macroscopic levels of plastic flow. The important things to remember are that some average plastic flow may occur at any level of loading and that in small regions of stress concentration the plastic strain is always larger than the average.

Thus, at least a partial role of mean stresses in the fatigue damage process is in the activation of minute stress concentrations. A considerably more complicated role of mean stresses is also plausible, and this will be explained in the following.

The concept is based on the possibility of at least a small amount of cyclic creep in microscopic regions. The factors affecting this creep process are the mean-stress level and the amount of cyclic plasticity. It should be recalled that the incompatibility of these two quantities allows only one of them to have a large value.

For an example, assume that the following conditions exist in a material cyclically loaded with a small amplitude but with a large tensile mean stress. The nominal stress amplitude is small, resulting in a small average cyclic plastic strain, as at point A in Fig. 101. Consider the state of the material at the tip region B of a small preexisting flaw or of a freshly nucleated fatigue crack. The large externally imposed tensile mean stress keeps this crack in an opened shape most of the time, so that the stress concentration is active. Consequently, the stress amplitude σ_B within the region B is much higher than the nominal value σ_a.

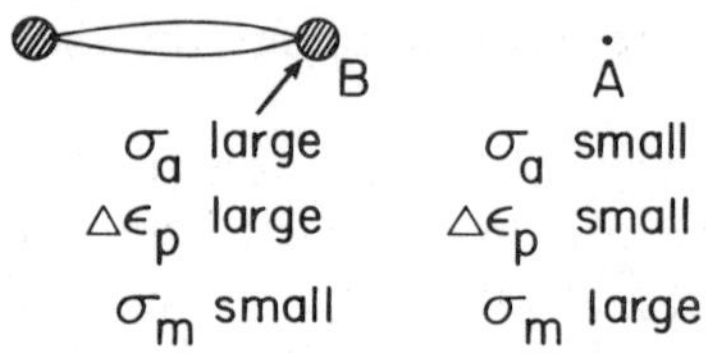

Fig. 101. Stresses and strains at different locations in a cracked member

The large stress σ_B causes considerable plastic flow, so that $\Delta\epsilon_{p,B}$ may be quite large compared to the nominal value $\Delta\epsilon_p$. In accordance with the incompatibility discussed earlier, the large cyclic plasticity at region B does not allow a large mean stress there. If the crack has just been nucleated, the large average σ_m at B will reduce gradually as the cyclic plasticity develops. For a preexisting flaw, the large σ_m at point B will also decrease to a level compatible with $\Delta\epsilon_{p,B}$. In turn, the stress concentration itself may be affected. The relaxation of mean stresses will be discussed in more detail in a later section.

Thus, the resultant state of the material at the crack tip comprises large cyclic stresses and plastic strains as well as a small mean stress. These contrast with the imposed nominal conditions far away from the crack tip shown in Fig. 101. The conditions at the crack tip can lead to at least a small amount of cyclic creep and thus slightly speed up the rate of fatigue damage accumulation.

It can be asked whether cyclic creep in the small volume of material is possible without the equal participation, in the creep deformation, of all the material. In other words, where can the material flow from the critical zone, when all of the surrounding material acts as a constraint? The answer is that, first of all, the total deformation involved is very small. The strains may be substantial, but the dimensions of the region are small. Also, using the example, while the crack tip region wants to creep in tension, the whole surrounding material is also experiencing a mean tension. Thus cyclic creep in tension at the crack tip could relieve, locally, part of the elastic tensile strain, that caused by the mean stress, in the neighborhood of the crack.

The effect of cyclic creep in a very small volume is like that of large-scale cyclic creep. In tension there is a gradual increase in tensile mean strain until the combined cyclic and monotonic plastic flows cause the material to fracture. For the small-scale creep this fracture is limited to the small dimension of the critical zone at the tip of the crack. Of course, the crack then grows in size and has a more critical effect on the fatigue life of the whole material piece.

Localized cyclic creep in compression, if it occurs at all, is less severe than its large-scale counterpart. The reason for this is that the material around the small critical zone furnishes the necessary constraint to prevent buckling. Thus the physical basis of the beneficial effect of compressive mean stresses becomes clear. The mean stress tends to close small cracks, rendering the geometric stress concentrations relatively inactive. At the same time, if cyclic creep in compression occurs, it can do so without the danger of buckling.

The possibility of locally enhanced cyclic creep becomes greater when either one or both of the following conditions exists. In the first, there is macroscopic cyclic creep of the whole material, so that the average mean strain is increasing. This requires large nominal loads. In the second, a relatively large part of the material is directly affected by the discontinuities. In that case only a small portion of the material could be considered to be nominally loaded even if these loads are small.

In summary, if we accept the possibility of some cyclic creep in the small regions of stress concentration, the effects of mean stresses in fatigue can be explained, qualitatively at least, as in the discussion above. Accurate quantitative analysis will be possible only when more becomes known about cyclic plastic behaviors at both the microscopic and macroscopic levels.

6.5 Residual Stresses

A particularly important and somewhat nebulous aspect of the mean stress problem arises in connection with residual stresses. By definition, a residual stress is one that exists in a material which is not loaded externally. Such a stress can truly be a residual when it is caused by external loads and remains after these loads are removed. The residual stresses are often created without the help of external forces. Electrochemical and solidification processes may lead to uneven internal stress distributions that exist without prior external loading.

The essential facts from the fatigue point of view are that most materials have some residual stresses in them and that a residual stress is a mean stress on which cyclic stresses are superimposed. It must be added that in the absence of external loads the internal forces must always balance each other. For example, it is not possible to have a piece of material with only compressive residual stress in it. Somewhere in the same piece there must be tensile residual stresses to maintain equilibrium.

The fatigue problem is immensely complicated by residual stresses. This is so even when the residual stresses are willfully introduced to improve the fatigue performance of a material. It is much more so in those common situations where complex residual stress patterns exist whose magnitudes and locations cannot be described to aid in the analysis. Many apparently inexplicable failures are caused by such conditions.

A common method of creating a beneficial residual stress pattern is by peening the surface. This introduces a compressive stress in the surface layers and delays the fatigue damage process. The compressive stress in the outer layers can be explained with the aid of the model in Fig. 102. Particles of material in row *A* are

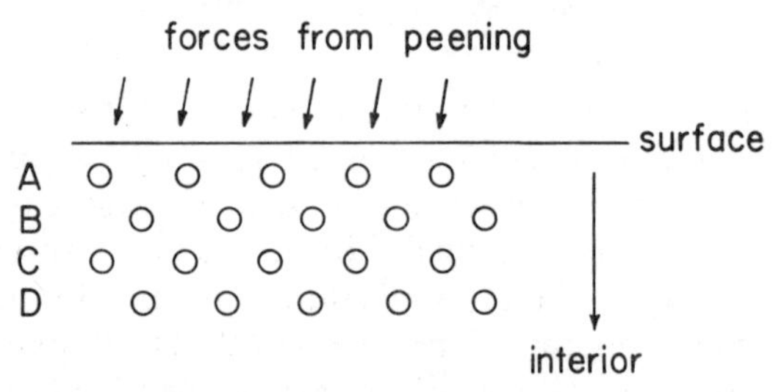

Fig. 102. Model of surface peening

at the surface, where the hammering forces are applied. Many of these particles are driven between the particles of row *B* immediately beneath them. Some of the original members of row *B* may be driven into row *C*. From row *C*, a still fewer number end up in row *D*, and so on toward the interior. In each row the particles were in equilibrium before the peening. The new arrivals forcefully occupy space that is normally shared by the original particles. It may be said that the extra members reduce the elbow room in each row, so that every particle feels squeezed by its neighbors, hence the compressive residual stress.

The resulting stress distribution is sketched in Fig. 103. The maximum stress is at the surface, where the interpenetration is most likely. Toward the interior, the compressive stresses decrease and eventually change to tensile stresses. These interior tensions are caused by the wedging action within the neighboring compressively stressed layers. The integral of the internal stresses over the entire cross section, of course, should be zero for equilibrium, in the absence of external forces.

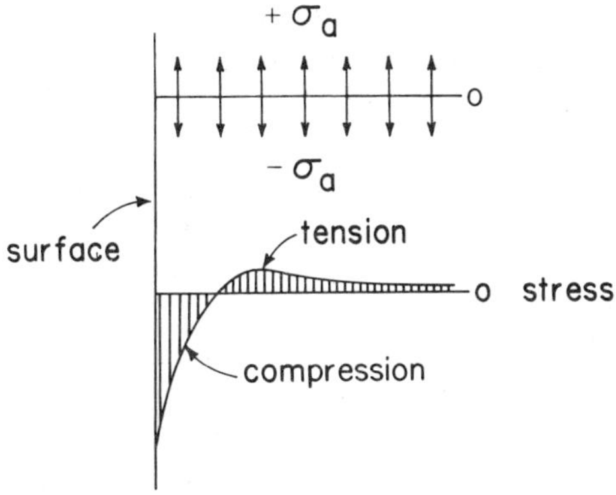

Fig. 103. Residual stress distribution; cyclic stress is superimposed on local residual stress

The pattern shown in Fig. 103 is quite beneficial for cyclic loading. For a uniformly imposed fully reversed loading, as in the top of Fig. 103, the resulting stresses are complex. The cyclic stresses are superimposed on the residual "local mean" stresses to make combinations that are relatively undesirable in the interior but desirable in the surface regions. The net result is favorable for two reasons. In the interior, the tensile stresses are spread over a large distance, so that for a given force needed to oppose the neighboring compression the peak tensile stress does not have to be high. The peak compressive stress at the surface is the highest of the residual stresses because it has been introduced forcibly and caused all the other stresses.

Another reason that makes the above stress distribution desirable is that in fatigue, the surfaces of materials are generally the weakest parts. It is at the surface that environmental effects are the strongest and the possibility of mechanical damage the greatest. Consequently, fatigue cracks generally initiate at the surface of a material. (Exceptions to this occur in materials that are not uniform in their mechanical properties or are not homogeneous.) A localized compressive mean stress at the surface is thus highly desirable for delaying the general fatigue-damage process.

The benefits from judiciously imparted compressive residual stresses can be impressive and economically important. Shot-peening of machine parts can significantly improve the fatigue life. For example, the high-cycle life of mild steel can be increased by a factor of about ten, and even higher for carburized steel [1]. At this point a word of caution must be raised about the expected benefits from compressive residual stresses. The residual stress pattern can drastically change during cycling, as will be seen in the next section. With these changes the expected benefits may not materialize. In general, it is well to deal cautiously with any problem involving residual stresses; even the best planned and executed project can turn into disaster as a result of cycle-dependent changes in the material.

6.6 Relaxation of Mean Stresses

One of the most important aspects of mean stresses in fatigue is that these stresses can with equal ease be either created *or* eliminated by plastic strain. In this respect, both monotonic and cyclic strains must be considered. The role of monotonic and quasi-static deformations in relation to mean stresses has been discussed with respect to the examples presented in Fig. 83. A possible role of cyclic deformations was shown in Fig. 84. The general incompatibility of large mean stresses and large cyclic plasticity was discussed with the aid of Figs. 85 through 88. All of these pertain to initial limitations on mean stresses in fatigue.

There is still another dependence of mean stresses on cyclic plastic strains. This is actually a cycle dependence, since the successive repetition of the plastic flows is a factor. The general observation is that under loading that produces significant cyclic plastic strains, the mean stress tends to become smaller as cycling progresses [51]. This is the case under strain control conditions, as in Fig. 104. The behavior is called cycle-dependent stress relaxation, and it is the counterpart of cyclic creep occurring under load control. In fact, the only way to maintain the mean stress at a true constant level is in load control, or when the strain limits are always adjusted as necessary to counteract the gradual decreases in mean stress.

The important facts about cyclic stress relaxation are that, regard-

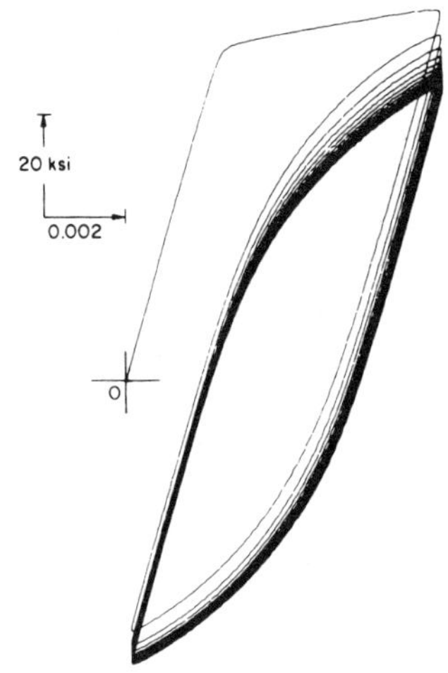

Fig. 104. Relaxation of a mean stress in SAE 1045 steel (from Landgraf, in *High Fatigue Resistance in Metals and Alloys*. Courtesy ASTM)

less of sign (tensile or compressive), the change in mean stress is normally a decrease in magnitude toward zero stress; that usually there is no crossover to the opposite sign; that the rate of change becomes smaller as the mean approaches zero; and that cyclic plastic strain facilitates the relaxation [80]. For a schematic representation of the relaxation process, assume a softening material that is cycled under strain control with a positive mean strain, as shown in Fig. 105. The material's response to the applied condition is the stress-

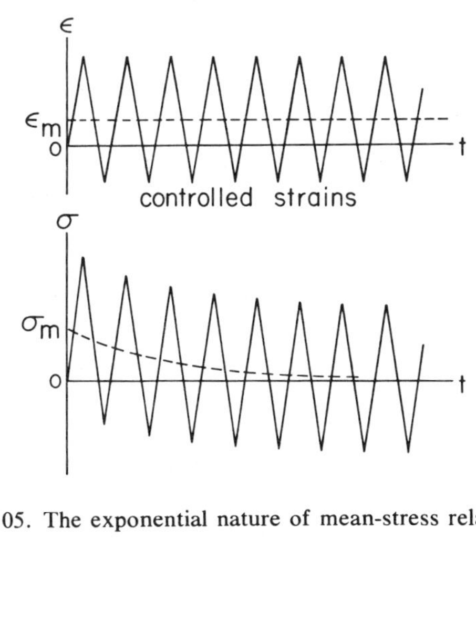

Fig. 105. The exponential nature of mean-stress relaxation

time plot on the same diagram. Assume that the initial mean stress is tensile, though it could be compressive as well, even for the same mean strain. The initial mean stress decreases according to the schematic, as cycling with plastic strains progresses.

An explanation of the relaxation process is illustrated by Fig. 106. The vertical walls are the strain limits, and at a particular

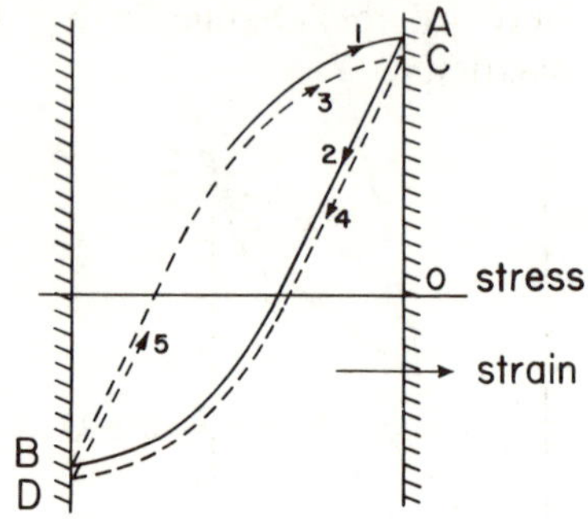

Fig. 106. Schematic of mean-stress relaxation for cyclically softening material

point in the cycling, the partial hysteresis loop is the solid line between *A* and *B*. The tensile mean stress is the stress corresponding to the center of the loop. We have assumed cycle-dependent softening, so that coming away from tip *B* the material cannot return to the tensile strain limit at point *A*. Instead, the loop will be reversed at a stress of *C* below point *A*. With a positive mean stress present, the tendency for plastic strain in compression is not enhanced, and consequently the loop shape from *C* to *D* is about the same as that from *A* to *B*. Thus there is a net change downward in the position (and σ_m) of the loop. On the following cycle, however, the decreased mean stress would be less effective in producing more plastic flow in tension than in compression. This way the rate of loop shift toward centering about zero stress is gradually decreasing.

For a cyclically hardening material the relaxation of the mean stress is even less obvious. The clue for the behavior is shown in Fig. 105. With cyclic softening, $\Delta\sigma$ is decreasing while σ_m approaches zero, but in principle, at least, these two quantities are not dependent on each other. The stress range could cyclically increase while σ_m is decreasing. The physical meaning of this is that hardening reduces the cyclic plastic strain and increases the stress range for the given total strain, but with a tensile mean stress the material yields more in the tension part of the cycle than in the compressive part. This preferential plastic strain would alter the loop shape in such a way (Fig. 107) that while tip *C* is above the previous tip *A*, the new tip *D* falls below the previous tip *B* by a larger amount. Consequently, the hardening material under a positive mean stress also has a net downward change in loop position. In this case, of course, the rate of change in mean stress is slowed not only by the change in σ_m itself but also by the decreasing cyclic plasticity.

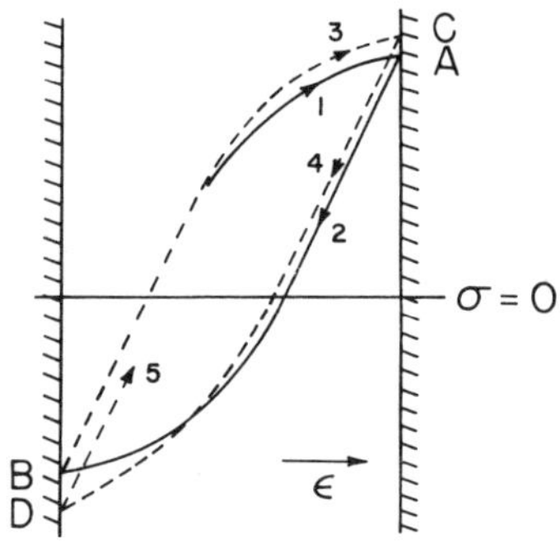

Fig. 107. Schematic of mean-stress relaxation for cyclically hardening material

The rate of relaxation depends on cyclic plastic strain, as shown schematically in Fig. 108. The decreasing functions are exponentials, even the nearly horizontal line. The basic philosophy of fatigue processes and their dependence on cyclic plastic strains should apply in this case, too. Therefore, some relaxation of mean stresses should go on, even under extremely small cyclic loads where the changes would not be easily observed. Mean stress relaxation has been observed at lives exceeding 10^6 cycles [51], but the presence of cracks toward the end of the life must be considered in evaluating

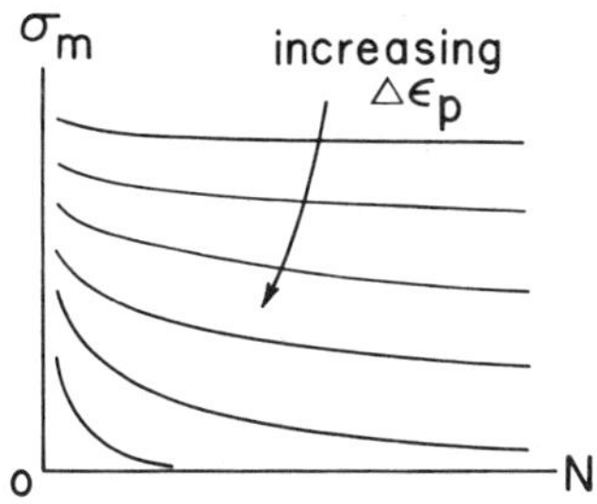

Fig. 108. The role of cyclic plasticity in mean-stress relaxation

such findings. At high cyclic plastic strains the curves may not stabilize, even in appearance, because fracture could intervene before the relaxation process is completed.

Total cyclic strains are measured relatively easily, but they are not the basic causes of cyclic relaxation. Nevertheless, a plot of average mean stress vs. total strain is useful in revealing the dependence of the relaxation process on cyclic plasticity. Figure 109 is such a plot, showing the general tendency in the behavior of four metals. Since the mean stress is not necessarily a constant throughout the life, the mean stress at half-life is selected as the representative value. Thus in the first half of the life the mean stresses are always above, and in the second half they are always below, the so-called average mean stress. In Fig. 109 the mean stress can be held constant when the loading is in the elastic range. Above the elastic range the mean stress drops at a striking rate with respect to increasing total strain range. In this region, plastic strains are becoming the dominant part of the total strains while the initial mean stresses are still high. The two are not compatible with each other, and this causes the rapid relaxation of the mean stress. As the mean value approaches zero, the rate of relaxation becomes negligible. It appears that in ductile metals a 3 percent total strain range can eliminate most of the mean stress by half the cyclic life [76].

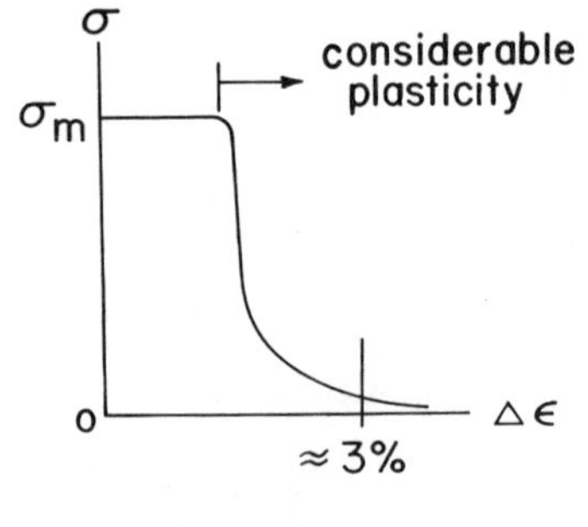

Fig. 109. Schematic of mean stresses at half-life under strain control

Such behavior has an important consequence in fatigue. Residual stresses, which are nothing more than unevenly distributed mean stresses, can "fade out" or "wash out" during cyclic deformations. Since the rate of fading out depends on plastic flows, large residual stresses cannot exist in low-cycle fatigue. This is also the reason why shot-peening is a useful technique only when the treated part is intended for long-life service.

There is still another, a somewhat subtle, problem in connection with mean-stress fadeout. This is the degree of relaxation that can occur in a relatively few overstraining cycles or even in large monotonic deformations. For instance, a machine part may be properly shot-peened to introduce beneficial compressive residual stresses in it and the service loads in the member kept small enough so that these residual stresses are stable for millions of cycles. It must be realized that at any point during the life of this member, a single cycle with large plastic flow could at once wipe out all of the residual stresses [51]. The minimum consequence of this overstraining event would be loss of the full benefit planned with the original compressive residuals. Of course, the overload might result in side effects of its own that could be complicating factors. These could include some direct exhaustion of life, and new residual stress of unpredictable signs or distributions.

The above discussion should make it clear that the most remote possibility of having any kind of mean stresses in a cyclically loaded member is to be taken very seriously. Such problems are always

Table 3. Strain-controlled step tests on 2024-T4 aluminum alloy; mean stresses

No. of blocks and sequence	$\Delta\epsilon_1/2$	n_1 per block	Mean stress, ksi	$\Delta\epsilon_2/2$	n_2 per block	Mean stress, ksi	Damage $\Sigma n_i/N_{fi}$
Tests with tensile mean stress:							
150 lo-hi	0.0054	17	4	0.0102	1	0	0.55
121 lo-hi	0.0046	17	7	0.0100	1	0	0.40
6 hi-lo	0.0102	50	0	0.0050	850	7	1.11
4 hi-lo	0.0110	50	0	0.0052	850	3	1.00
3 hi-lo	0.0100	50	0	0.0034	5000	17	0.40
						average =	0.69
Tests with compressive mean stress:							
6 hi-lo	0.0104	50	0	0.0050	850	−6	1.14
6 hi-lo	0.0102	50	0	0.0035	5000	−17	1.10
8 hi-lo	0.0097	50	0	0.0049	850	−15	1.46
5 lo-hi	0.0051	850	−9	0.0099	50	0	1.08
11 hi-lo	0.0100	50	0	0.0036	500	−30	2.20
8 lo-hi	0.0034	500	−29	0.0095	50	0	1.18
						average =	1.36

Source: Topper, Sandor, and Morrow, "Cumulative Fatigue Damage Under Cyclic Strain Control." Courtesy ASTM

Note: Mean stresses are "steady state" values at the smaller strain range. Damage summation is based on completely reversed strain-life data with no attempt made to account for mean stress.

difficult to handle but especially so under complex loading patterns. The simplest cumulative damage law, the Palmgren-Miner linear relationship, can lead to large errors in the prediction of life. Table 3 shows several results of damage summation for relatively simple multiple-level tests.

Deviations from the ideal sum of unity often exceed those shown in the table [67, 76]. It is generally possible to find empirical relationships that make corrections for the summation errors caused by mean stresses [75]. The empirical expressions have no physical bases of universal significance, however. This way every material must be laboriously evaluated for each specific loading condition, obviously an unfortunate necessity. Better methods may be developed when life correlations become available for any level and sequence of cyclic plastic strains, and when the effect of mean stresses in changing the cyclic damage rate is well understood.

7

The Notch Problem

Most of the discussion up to this point has been based on the assumption that the material under fatigue loading was relatively smooth geometrically. This assumption excludes cases in which section dimensions change abruptly because of discontinuity of any kind. There have been a few remarks made already in the preceding sections about the role of small flaws in fatigue. In general, the notch problem is extremely important in fatigue, regardless of the scale of the discontinuity.

Notches create complications in fatigue because they make the stress distributions uneven and are involved in the development of residual stresses. In the simplest case of an ideal fully elastic material shaped symmetrically and under uniform static external load, the stress distribution becomes as shown in Fig. 110. Here σ is the applied stress and σ_n is the nominal, or average, stress at the minimum section. The only value in considering σ_n is that it gives a vague idea of the adequacy of the area A_2. These two stresses are of course linearly related to each other by the ratio of the areas, and $\sigma_n = (A_1/A_2)\sigma$. The actual stress distribution has peaks at the stress concentrations that are considerably greater than σ_n.

The actual material behavior at the root of the notch is complicated by plastic yielding [10] under the highest stresses. This yielding makes the analysis more difficult than in the case of completely elastic deformations, but it is often beneficial in practice (except in plastically induced fracture) because of the resulting blunting of cracks [72]. Even though it is best to assume some plastic flow in any notched member under load, the idealized elastic behavior is useful in analysis because it can be predicted with relative ease on the basis of notch geometry.

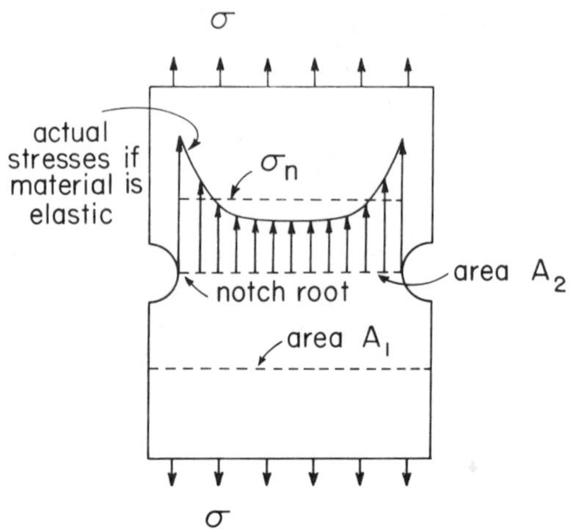

Fig. 110. Stress distribution in a notched member

The theoretical severity of a notch is described using the stress concentration factor K_t. This is defined as the ratio of the maximum local stress to the nominal stress in the region. The local stress may be determined mathematically, or by experimental stress analysis. It is important to remember that in the factor K_t the local stress is based on idealized completely elastic stress-strain response. In practice such perfect behavior is seldom approached.

In actual deformations at a notch tip there is often considerable plastic flow, which alters the fatigue strength reduction of the member. This reduction is less severe than that which could be anticipated on the basis of the theoretical stress-concentration factor. In order to describe the effects of notches in fatigue, a so-called fatigue strength reduction factor, K_f, has been introduced and defined as the following ratio at a given life:

$$K_f = \frac{\text{fatigue strength without stress concentration}}{\text{fatigue strength with stress concentration}} \tag{47}$$

This factor depends on the material, since the ability to plastically deform at the notch root affects the life for a given geometry. Nevertheless K_f is not a material property, because it depends not only on the material but on specimen size and shape, state of stress, and fatigue life range as well.

The factors K_t and K_f may or may not be the same. Their relative magnitudes indicate the degree of notch sensitivity of a material in fatigue. By a commonly accepted definition, the notch sensitivity factor q is given as

$$q = \frac{K_f - 1}{K_t - 1} \tag{48}$$

This factor may have any value between two extreme possibilities. Full notch sensitivity is obtained when $K_f = K_t$ so that $q = 1$. Notch insensitivity, $q = 0$, results from $K_f = 1$. Table 4 illustrates the notch behavior at a given life for several metals. The values shown are not directly suitable for quantitative design [55], because they are not material properties but rather depend on a number of factors, as mentioned previously.

Table 4. Stress-concentration factors, K_t, fatigue-notch factors, K_f, and notch-sensitivity indices, q, for various alloys in rotating bending

Alloy	K_t	K_f	q
Aluminum 2024-0	1.6	1.0	0
Magnesium AZ80-A	1.6	1.1	0.16
Stainless steel, type 18-8	1.6	1.0	0
Structural steel (BHN = 120)	1.6	1.3	0.5
Hardened steel (BHN = 200)	1.6	1.6	1.0
Gray cast iron	1.6	1.0	0
Bronze forging	1.6	1.0	0
Aluminum 7075-T73	6.7	1.8	0.13
Titanium 6A1-4V	3.5	2.8	0.72

Source: Osgood, *Fatigue Design.* Courtesy John Wiley & Sons, Inc.

It can be generalized that very hard, brittle materials are extremely sensitive to notches because of their limited ability to deform plastically. In these materials, the stress concentration can easily raise the peak elastic stress to the level of inherent cohesive strength between atoms, and cause fracture [72]. Ductile materials are less sensitive to notches, because of plastic flow that prevents forever-rising stresses. From the notch sensitivity point of view, cast irons are rather interesting, because they are relatively brittle and yet insensitive to notches. The reason for this is in the structure of the metal. Cast irons are so full of holes in the form of graphite and actual flaws that machined discontinuities do not make much difference in the overall behavior of the metal.

For more detailed background information on stress and fatigue concentration factors and related problems the reader should consult references 16, 19, 53, 55, and 58.

7.1 Fatigue Prediction of Notched Members

One of the important aspects of the study of fatigue is a never-ending search for simple experimental or analytical methods that can be used for the prediction of fatigue behavior in relatively complex situations. A good example of this approach was the use of monotonic test data in attempting to predict the fatigue life. Even for the simplest and most ideal specimen geometry and loading pattern such a prediction is a substantial step forward. In a similar manner we can conceive the prediction of fatigue behavior in a notched member on the basis of known cyclic behavior in a simpler smooth specimen. Recent advances in this area have produced interesting results and raise the hope of having fairly useful methods in the future [82]. The basis of the prediction is Neuber's approach, which takes into account the possibility of nonlinear deformations at the root of the notch. This can be described in the following way.

The theoretical stress concentration factor, K_t, is valid only in the elastic range, as in part *OA* of the stress-strain curve in Fig. 111. At high enough average stresses the notch root yields, and

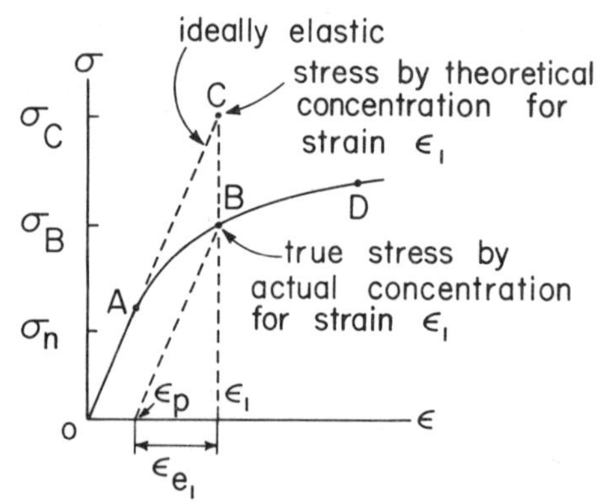

Fig. 111. True stress concentration

the maximum local stress increases only a small amount even for a large increase in nominal stress at the section. Thus the stress at the notch root for a given strain ϵ_1 is only σ_B in Fig. 111, and not the imaginary elastic stress σ_C. The average stress, σ_n, may still be in the elastic range, although this is not a necessary condition.

In this nonlinear situation a stress concentration can still be defined, but it is not K_t anymore. We can establish a true stress concentration factor, K_σ, in terms of the true local peak stress, σ_B, and the nominal stress, σ_n. Of course σ_n is a somewhat fictitious stress since, as shown in Fig. 110, it is the true stress at only one or a few points. Nevertheless it is useful because it can be easily calculated. Therefore the true stress concentration for a peak strain ϵ_1 is

$$K_\sigma = \frac{\sigma_B}{\sigma_n} = \frac{\sigma_{\text{true}}}{\sigma_n} \tag{49}$$

It is obvious that K_σ is not a constant; it depends on the applied load, and it is a cycle-dependent quantity as much as the stress-strain curve itself is.

An interesting way to look at the notch problem is that, while stresses are less concentrated than predicted from the ideal behavior, plastic strains are concentrated mostly at the notch root. Consequently, we can define a plastic strain concentration factor, K_p, as

$$K_p = \frac{\epsilon_{p\,\text{local}}}{\epsilon_{p\,\text{ave}}} \tag{50}$$

where $\epsilon_{p\,\text{ave}}$ is the nominal plastic strain (caused by the nominal stress).

The most common way of treating the concentration of strains is by dealing with total strains instead of plastic strains. The method is similar to that above and the strain concentration factor is

$$K_\epsilon = \frac{\epsilon_{\text{local}}}{\epsilon_n} \tag{51}$$

where ϵ_n is the nominal total strain.

The total-strain approach is often adequate in practice, but strictly speaking it is incorrect, because it includes the concentration of the elastic strain component, which is essentially a representation of the stress concentration. The reason that Eq. 51 can generally be used instead of Eq. 50 is that most ductile metals have gently sloping stress-strain curves in the plastic region, as in Fig. 111. Thus the elastic strain components at the notch root (points B or D) are not much higher than the respective component of ϵ_n far from the notch. This way a large concentration of total strain is approximately a measure of the concentration of plastic strain.

Neuber developed a mathematical expression of the stress and strain concentrations at the notch root [77, 82]. Formally,

$$K_t = (K_\sigma K_\epsilon)^{1/2} \tag{52}$$

In words, the theoretical stress concentration factor, a constant, is equal to the geometric mean of the true stress and true total strain concentration factors. Equation 52 is exact when the notch root remains elastic. In this case, any concentration of stress is accompanied by the same concentration of strain, which can be calculated using Hooke's law. Thus $K_\sigma = K_\epsilon$ and from Eq. 52, $K_t = K_\sigma$. The validity of Neuber's expression is not necessarily obvious when yielding takes place at the notch root. However, it was shown that K_σ decreases and K_ϵ increases in nonlinear deformation, so that their product could be a constant. Accurate analysis of notched members in fatigue requires more research in the area of concentrated stresses and strains at notches.

For applying Neuber's approach in fatigue analysis, K_f is used instead of K_t, and the other two factors are written in terms of cyclic stress and strain quantities [77, 82]:

$$K_f = \left(\frac{\Delta\sigma}{\Delta S}\frac{\Delta\epsilon}{\Delta e}\right)^{1/2} \tag{53}$$

where $\Delta\sigma$ and $\Delta\epsilon$ are the true stress and strain ranges at the notch root, and ΔS and Δe are the nominal (average) stress and strain ranges on the cross section containing the notch. Equation 53 is rewritten, for convenience, as

$$K_f\,(\Delta S\,\Delta e\,E)^{1/2} = (\Delta\sigma\,\Delta\epsilon\,E)^{1/2} \tag{54}$$

The elastic modulus is put in to give both sides simple units of stress.

The importance of Eq. 54 lies in the way the various quantities are grouped. It must be remembered that $\Delta\sigma$ and $\Delta\epsilon$ are true quantities at the notch root, where they would be very difficult to measure. At the same time, ΔS and Δe are nominal values that can be easily computed from the external load and the stress-strain curve. It may be noted that in most engineering problems ΔS is in the elastic range and $\Delta\sigma$ is in the inelastic range.

Equation 54 can be used to predict the fatigue life of a notched member if the fatigue behavior of the same material is known from tests on smooth specimens in which $\Delta\sigma$ and $\Delta\epsilon$ can be measured. Detectable cracks are expected in both kinds of specimens at about the same life if $K_f\,(\Delta S\,\Delta e\,E)^{1/2}$ for the notched specimen is equal to $(\Delta\sigma\,\Delta\epsilon\,E)^{1/2}$ for the smooth specimen. Use of the method for two aluminum alloys is shown in Fig. 112. Part *a* of

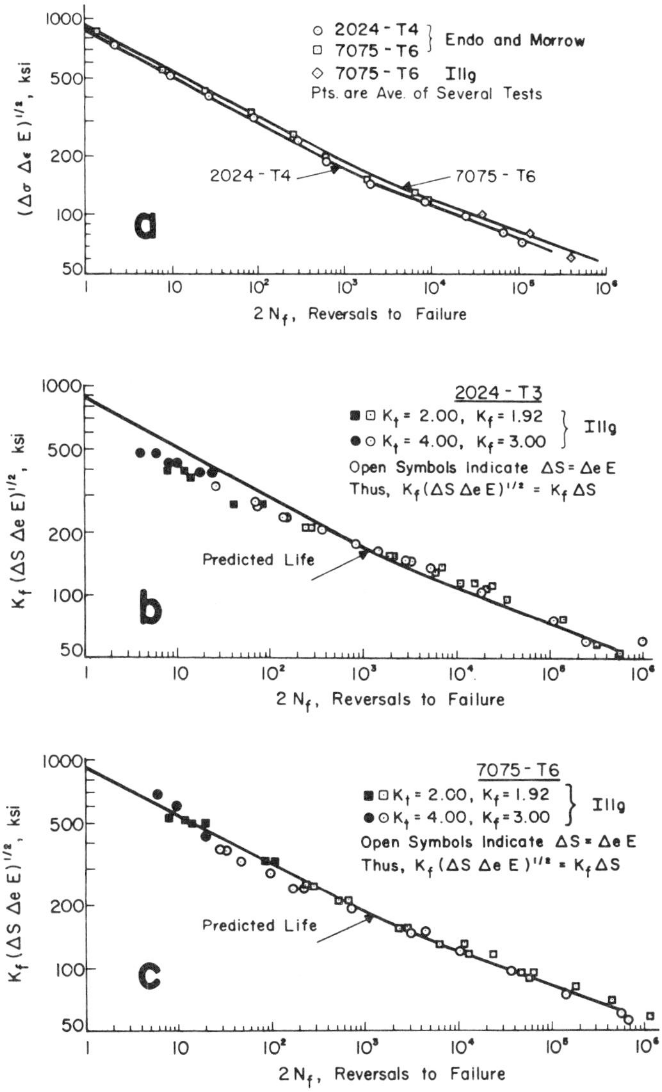

Fig. 112. Prediction of fatigue life of notched members: (a) fatigue data from smooth specimens in a form suitable for prediction of life of a notched member; (b and c) comparison of data from notched members with the predicted life (from Topper, Wetzel, and Morrow, "Neuber's Rule Applied to Fatigue of Notched Specimens." Courtesy ASTM)

the figure is a master plot of smooth specimen data vs. life. These plots are assumed to represent the notched specimen behavior on the basis of Eq. 54. For clarity, the first half of Eq. 54 is plotted on separate diagrams (parts *b* and *c*) as a function of life for notched specimens. The respective coordinates of the separate parts in Fig. 112 are exactly the same in units and in scale. Thus for a particular cyclic loading, the value of $K_f\,(\Delta S\,\Delta e\,E)^{1/2}$ can be calculated and entered on the ordinate. The corresponding point on the master curve gives the life of the notched member. If the nominal load ΔS is large enough to cause nominal yielding, Δe must be measured or estimated from the appropriate cyclic stress-strain curve. At small loads the nominal behavior is elastic and $\Delta S = \Delta e\,E$.

Figure 112 shows that for the two examples the largest error in predicted life was about a factor of two. This is good agreement, considering the newness of the art and the possible scatter in results. One limitation in the method is that while it predicts first cracking in the notch region, in some cases it may take considerably longer to propagate the crack through the remainder of the material. A more serious limitation of the above approach is that as yet it cannot account for mean and residual stresses. In such cases the predicted lives may be in error by more than two orders of magnitude [77]. The possible relaxation of mean stresses at the notch root have not yet been considered in fatigue analyses.

On the basis of the necessary emphasis on cyclic plastic strains, it appears that the given form of Neuber's concept may be adequate for some life prediction but that it is not the most precise form for damage analysis. It is conceivable that a fatigue concentration factor of accurate physical basis could rely entirely on the concentration of plastic strains. In that case the concentration of stresses may be considered superfluous, since these stresses cause fatigue damage only to the extent that they cause cyclic plastic strains. A fatigue concentration factor of general applicability may, of course, also include mean stresses or the rate of cyclic creep (monotonic plastic deformation), to account for the acceleration of damage accumulation. It is obvious that much further research is needed in this important area of materials fatigue.

7.2 The Control Condition at a Notch

From the very beginning of this discussion of cyclic stress and strain two factors have been emphasized. One is the externally imposed loading control condition and the other is the characteristic material stress or strain response to the applied condition. In simple cases, the first of these is automatically given by the specification of the problem, and the response is relatively difficult to obtain. The notch problem reverses our view of these two factors somewhat. The material response is the easiest to obtain, since it should be the same as in the fatigue of smooth specimens for the given material under appropriately specified conditions. The actual control condition at the root of the notch is more problematic, because different portions of the stress-strain curve apply for the notch region and for the material away from the discontinuity. A precise knowledge of the imposed condition in the critical region is necessary for an analysis of the fatigue behavior of the whole structure.

One simplified way to describe the events at the notch root is by imagining that we are dealing with a composite material of two different components. The critical zone at the notch is one of these components, and everything else is a different material. In this way a complex pattern of possible situations becomes evident. The basic schematic of the material combination under external load is shown in Fig. 113. Generally, the notch directly affects only

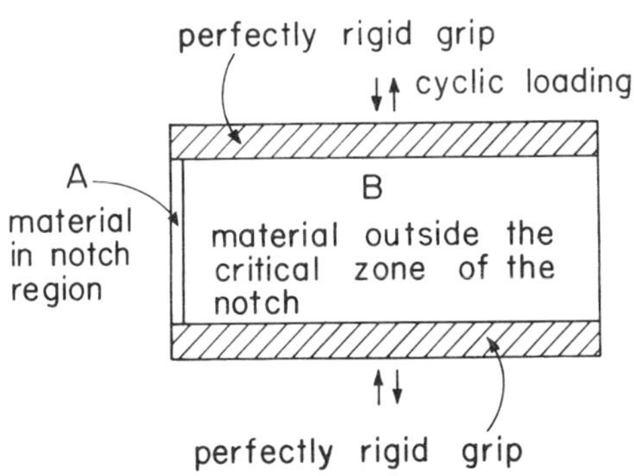

Fig. 113. Model for the behavior of notched members

a small volume of material, region *A* in the figure. The remaining portion *B* is the bulk of the structure. The given combination is assumed to be clamped between perfectly rigid plates that always move parallel to each other under any kind of external load.

Two extremes of cyclic loading are considered, load control and strain control. In load control, cyclic force limits are applied to the grips regardless of the deformation response of the material between the grips. Obviously the component *B* must carry most of the load, and before cracking it does not matter much what the small amount of material *A* does in contributing to the load resistance. Under strain or deformation control, the relative parallel position of the grip plates is controlled, regardless of how much force is generated in the material between the grips. Again, the presence of material *A* does not affect the behavior of *B,* since they are experiencing equal deformation between the imposed limits. The conclusion is that for parallel material combinations under any kind of cyclic loading, the smaller component does not appreciably affect the larger component's behavior if the volume ratio is large.

The effect of the larger component on the smaller one is more complicated. When the external loading is controlled strain the situation is the same as above. Each component experiences the same externally enforced deformation entirely independently of the other. Under external load control the small volume of material comes under the control of the larger. The reasons and results are examined in the following.

One extreme of the possible behaviors is severe cycle-dependent softening within the small region *A* (the notch root). Even though cyclic force is controlled externally, a runaway situation does not develop in region *A* because it is firmly attached to *B.* In this case *A* carries a progressively decreasing share of the total load. Such localized softening does not noticeably impair the overall strength of the piece, because region *A* is small. The important thing to realize is that *A* is experiencing mainly strain control while *A* and *B* together are under cyclic forces. The reason is that *B* carries essentially the full load, quite frequently by elastic deformations. These elastic strains become the control function for the attached small volume *A.* In other words, the main portion of the material combination responds by a certain strain to the imposed load, and the small part simply rides along.

Another extreme possibility is severe cycle-dependent hardening in the notch region *A.* From the point of view of localized control, this is not different from the softening situation. The small volume now develops a larger force per unit strain than a section of same size does within the large portion *B.* Most of the external load still must be carried by *B,* however, so that again deformations in *A* are determined by the stress-strain response of the major component. In a sense, the improved load-carrying ability of *A* is ignored by the material in the larger neighboring area. Thus it appears that the approximate control condition at a notch root is that of cyclic total strain regardless of the externally imposed condition. We must speak of approximate control because in reality parts *A* and *B* are often of the same material and there is no sharp dividing line between them. Even if there were such a line, the control condition in region *A* would be somewhat complicated by the fact that it is transferred into *A* through the side attached to region *B*.

The question may be raised whether it is important to consider the control condition at the notch root at all. Some of the behavior at the root is caused by the comparatively small amount of material affected in its immediate neighborhood. It may appear that, in turn, this small region should not appreciably affect the fatigue behavior of the whole piece. In reality, the small region around the notch is highly critical. Regardless of the local control condition, the plastic strains are largest at the notch. The locally high damage rate causes cracks to start at the notch root. Once started, these cracks can propagate relatively easily through the whole part. The local control condition is important because of the cycle-dependent changes occurring at the tip of the notch. For example, if there is a mean stress in the critical zone, it should tend to relax toward zero under the local strain-control condition. Inasmuch as the local control is not necessarily a perfect strain control (since not all of *A* is directly attached to *B* in Fig. 113), a small amount of cyclic creep could occur under the mean stress in the critical region. An accurate view of the notch problem will be possible only when the cyclic stress-strain behavior in the smallest root regions comes to be understood well.

7.3 Notches and Residual Stresses

Notches should always be considered with respect to their important contributions in generating residual stresses. These stresses may be caused by a single large load on the member, or by an abrupt transfer from large to small cyclic loads [26, 67]. Of course, these residual stresses can be eliminated just as easily as they are created, if there is substantial cyclic plasticity. In this section the emphasis will be on the peculiar aspects of the notch problem with respect to the creation of residual stresses.

The behavior appears unusual at first because it seems to oppose the general findings in fatigue. One tends to develop a feeling that anything of a tensile nature is detrimental in fatigue, while compression must be generally beneficial. Examples that strengthen such concepts include tensile mean stresses, cyclic creep in tension, and compressive residual stresses. The unexpected observations are that notched members overloaded in tension and subsequently cycled under small, fully reversed amplitudes have increased fatigue lives.

The reverse is also true, in that a large compressive load reduces the life at following smaller amplitudes.

It is found upon closer examination that the observed fatigue behaviors are not unusual in their basic mechanisms. A tensile overload is beneficial because it creates a compressive residual stress at the critical region of the notch root. Compressive overloads are harmful because they leave tensile residual stresses. These help in explaining the fatigue behaviors as expected. However, the mechanism of generating the residual stress by an overload remains difficult to understand. The following discussion of this subject is for a tensile overload, but the principles also apply for compressive loading.

Figure 114 shows the internal stress distribution for a notched

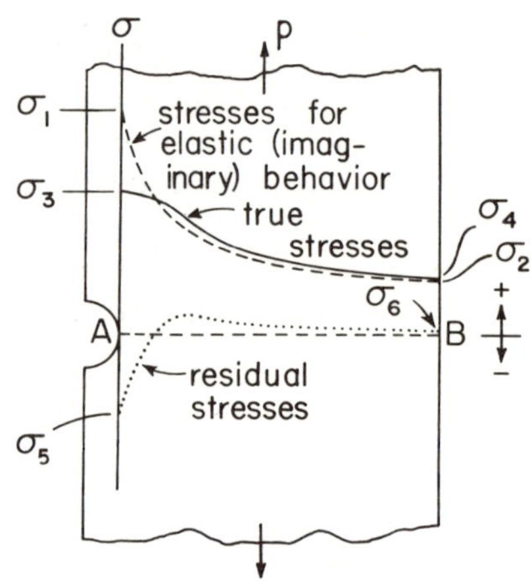

Fig. 114. Residual stresses created by an overload

member. Let us assume that the broken line is the imaginary distribution of true stresses in the perfectly elastic case. The solid line represents real stresses when yielding occurs at the notch. Away from the notch region the real stresses are higher than the imaginary ones, because the total resisting force (the integral of the stresses) must be the same for the two distributions. The idealized stress pattern is important in discussing the notch problem because the unloading part of the overload event is essentially an elastic process.

Unloading from the fully elastic curve is a reversal of the loading process. Every material point within the cross section *A-B* is able to return to zero stress when the external force is completely removed. The rate of unloading at a given point is the same as the loading rate. For instance, the removal of a unit load (external) lowers the stress at point *A* the same amount as the stress would be increased by a unit load applied to the member (elastic behavior).

Inelastic deformations upon loading cause an uneven stress distribution after subsequent external unloading. The reason for this is that the local rate of stress increase may be governed not only by the linear stress-strain behavior and the stress concentration factor but also by plastic flow. In contrast to this, plastic strain is not significant in unloading, or even in a following small reversed loading. According to Fig. 114, for example, the maximum stress at point *A* may be limited to σ_3 by plastic flow. When the external load is removed, the stresses at *A* decrease in an elastic fashion in such a way that the local change in stress tends to be all of the magnitude of σ_1. Measuring the magnitude of σ_1 downward from σ_3 gives the final stress σ_5, which is in compression. The same procedure can be followed for all points of the cross section, keeping in mind that the elastic unloading is to start from the curve of true stresses. The resulting dotted line in Fig. 114 is the plot of residual stresses present after unloading. Of course it is entirely possible that the magnitude of σ_5 itself is high enough to cause some yielding in compression.

Another way to look at the above problem is to consider the stress-strain curve of the material, as illustrated in Fig. 115. Various

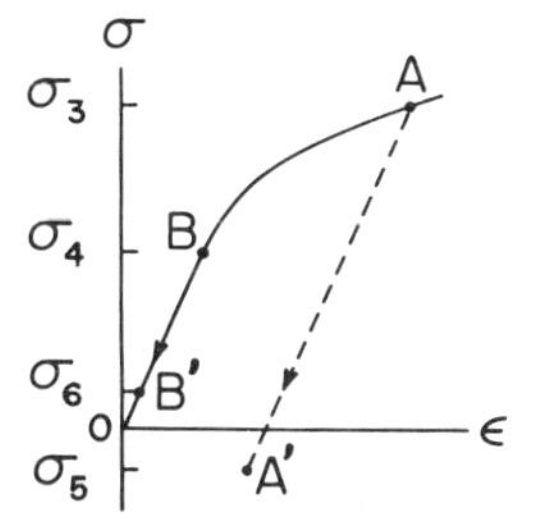

Fig. 115. Analysis of stresses in overloaded notched member; symbols agree with those in Fig. 114

points in the cross section *A-B* (Fig. 114) are strained differently because of the stress concentration. Point *B* still may be in the elastic region, while point *A* has plastically deformed. Regardless of the position of any of these points on the stress-strain curve, unloading will cause the stress to decrease along elastic lines. In the absence of an external load, the material at all points is striving to reach a state of zero stress. The stress concentration may be fully effective during unloading, and the notch root (*A*) changes stress much more rapidly than any other point does. Thus, even though point *A* may be starting from the highest stress level, it reaches zero stress and then goes into compression before point *B* (and many others) have returned to zero stress. The resulting arrangement of residual stresses can be in stable equilibrium because the tensile and compressive forces cancel each other internally. It is true, of course, that for a metal with a large amount of sudden yield the notch stress is not necessarily the highest when the member is first loaded. In this case the unloading may produce a more extreme reversed stress at the notch root.

7.4 Single and Multiple Overloads

The preceding discussion of overloads refers to an external load that is large enough to cause yielding in at least a small volume of material at the notch root. The schematic representation of a single overload is shown in Fig. 116. Whether the large load is

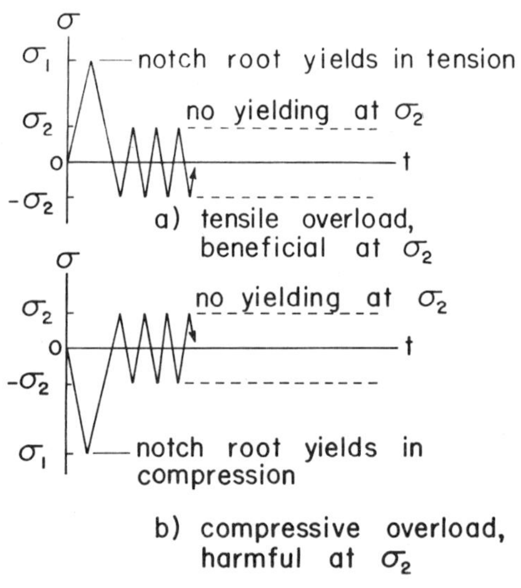

Fig. 116. The effect of tensile and compressive overloads

harmful or not with respect to residual stresses is determined not only by the magnitude and direction (sign) of the large load but also by the subsequently imposed smaller loads. Obviously, if the small loads themselves are large enough to cause substantial cyclic plastic flow, the residual stress can rapidly relax toward zero stress. Similarly, the residual stresses may never appear if the transfer from the large load to the small ones is gradual, as in Fig. 117. In this case the secondary stress amplitude σ_2 may be small enough

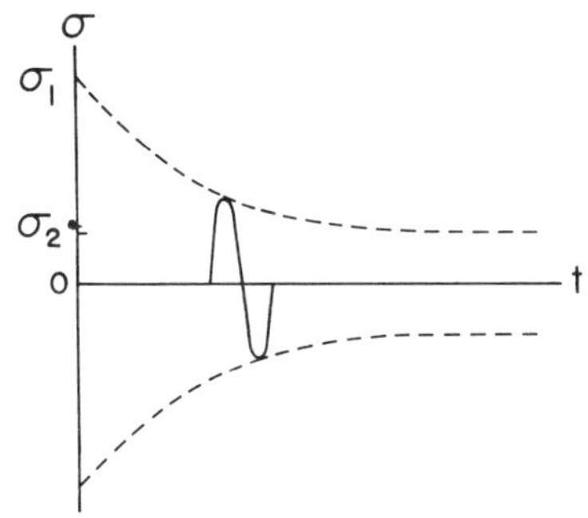

Fig. 117. Transition from large to small loads without leaving significant residual stresses

to tolerate a large residual mean stress, but the cyclic plastic strains occurring during transfer prevent the development of the residual stresses.

The general problem of multiple overloads is the same as that of a single large load, as far as the generation of residual stresses is concerned. Attention should always be concentrated on the last half-cycle of the large loads before the transfer is made to smaller amplitudes. A typical example of this is shown in Fig. 118, which

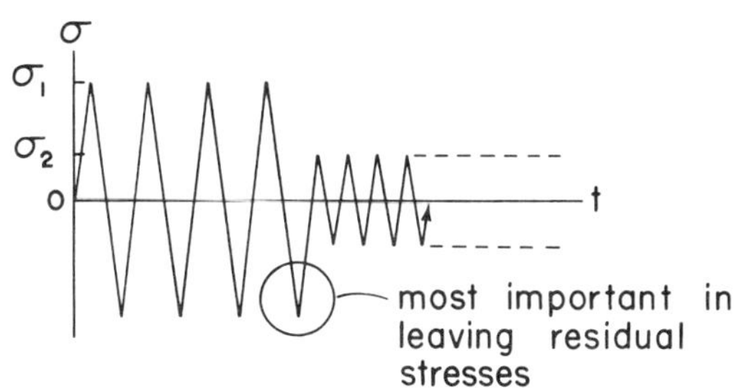

Fig. 118. Residual stress after multiple overloads

gives the same residual stresses as the single overload in Fig. 116*b*. It should be noted that the stresses shown in Figs. 116 through 118 are the externally imposed loads on the notched member and not the true local stresses.

The practical significance of residual stresses by occasional large loads cannot be overemphasized. Most load-carrying members have notches in them, and it is likely that most of them experience at least a few large loads during their service lives. The following examples illustrate some possible situations in complex structures.

In the early days of aviation history many airplanes were tested before their first flight in a manner sketched in Fig. 119 (from a photograph of a small airplane during its final checkout at the factory). Each wing was loaded by the weight of five men, supposedly to make sure that the wing would not fall off during flight. Let us consider this "test" from the fatigue point of view.

Fig. 119. Early method of preflight testing

The important factors are the relative deformations of the wing during the loading test and during later flight. Figure 120 shows

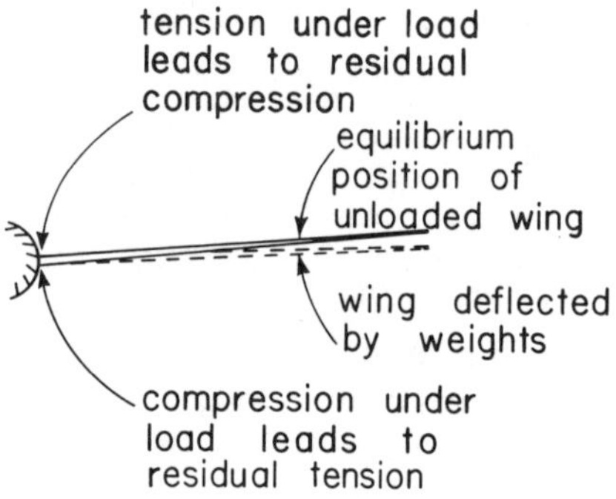

Fig. 120. Wing before and after test-loading

the stress conditions at the critical wing-root region during the artificial loading and just after this load is removed. Under the weight, the wing bends downward, and this may overload the material at any stress concentration at the root of the wing. The loading produces tension at the top of the wing and compression at the bottom. When the extra weight is removed the wing rebounds to nearly its original position of equilibrium. This may leave residual tensile stresses at the bottom and compressive ones at the top, if the overload is large enough. These stresses are generated in the same way as in the axial loading shown in Fig. 114.

During flight, the same wing experiences a different loading, as shown in Fig. 121. The lift forces bend the wing upward from its ground equilibrium position, and there are cyclic deflections

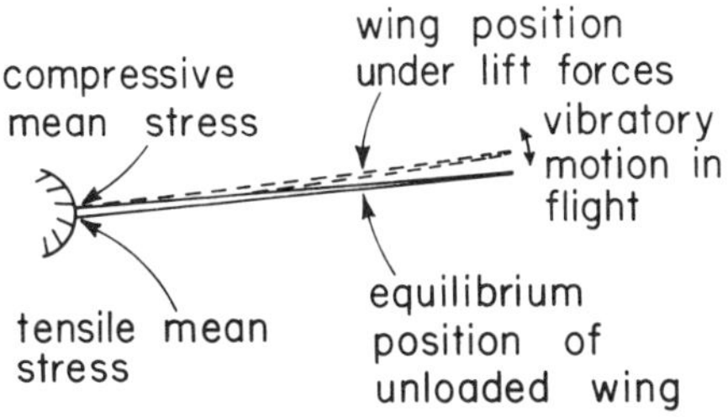

Fig. 121. Wing in flight

superimposed on this flight equilibrium position. Thus the root of the wing in flight is cyclically loaded, with a tensile mean stress on the bottom and a compressive one on the top. It is seen that the residual stresses created during the test (Fig. 119) are added to the local mean stresses that are present during flight. The combined stresses are obviously highly detrimental to the fatigue life of the wing structure.

Interestingly enough, modern airplanes are still subjected to the kind of loading shown in Fig. 119. When a plane is involved in a hard landing, the rapid downward motion of the fuselage is stopped by the impact through the landing gear, while the wings tend to continue their motion toward the ground. The resulting residual stresses at the wing roots are unfavorable, as those in Fig. 120 are. This kind of loading may be particularly undesirable, because even though the residual stresses might have a tendency to fade out later, they would be periodically reintroduced at the critical locations.

The overload by landing impact is different when the landing gear is attached to the wings, as shown in Fig. 122, instead of

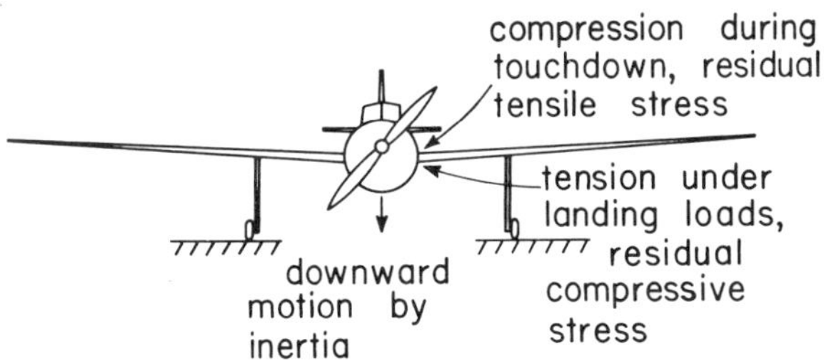

Fig. 122. The effect of landing, at some critical points of an airplane

to the fuselage. Considering the root of the wing only, a hard landing may be relatively beneficial. In this case the wing tends to bend upward from the fuselage during touchdown. This overload leaves residual stresses at the wing root that are favorable during the subsequent aerodynamic loading.

The above examples show that the generation of various residual stresses is always a possibility when complex structures are subjected to cyclic loads whose amplitudes are not constant. These stresses should be at least qualitatively evaluated for their signs and locations to determine the places of interest for further quantitative analysis. Such investigations are necessary to assure the economical production of reliable structures.

8

Unusual Fatigue Behaviors

The subject of material's fatigue is so broad that the description of all of its known aspects would require many volumes. The discussions in this text have been based on the fundamental mechanics (stress-strain) concepts of fatigue processes. These concepts of cyclic stress and strain are most useful in dealing with fatigue because they are conveniently situated in the middle of the scale of phenomenological observations. Stress and strain measurements in the laboratory are more desirable in fatigue analysis than, for example, the full-scale tests of airplanes. At the same time, the real cause of fatigue damage must be sought in events at the atomic level. Such investigations are extremely useful but difficult to accomplish and to relate to the large-scale behavior. Cyclic stress-strain patterns are not only indicators of the processes at the atomic level but also serve as a bridge between the small- and large-scale phenomena.

In this section, ten examples will be given to illustrate the versatility of the mechanics approach to the fatigue problem. The term *unusual behaviors* is not strictly correct, since everything in fatigue could be considered somewhat unusual. Many of these examples would fit well in other chapters, in fact. They are grouped here to illustrate that it is dangerous to take anything for granted in fatigue without having experimental evidence applicable to many aspects of the problem under study. They are also intended to show that concepts of cyclic stress and strain are strong bases for understanding complex fatigue situations.

8.1 The Shape of the Hysteresis Loop

Most properly obtained hysteresis loops have sharp tips and similar paths between the tips, as shown in Fig. 39. On rare occasions the recorded loop is blunted, as shown in Fig. 123. In this case

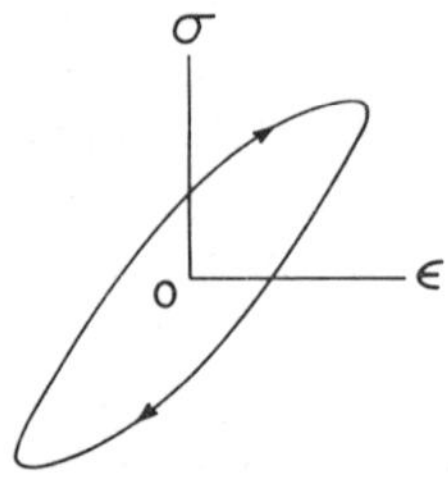

Fig. 123. Distorted hysteresis loop caused by stress relaxation

the first impulse should be to check the instrumentation. It is possible, however, that this is truly the behavior of the material. Such a loop may be obtained when the cycling is slow, the temperature is a substantial fraction of the melting temperature, and the stress amplitude is high enough to cause creep. In this case the material has time to relax the peak stress toward zero stress before the load reversal occurs. Immediately after the reversal, the stress is still high enough to cause a little further relaxation, which distorts the unloading path. The total plastic strain per cycle has to include creep strains, since relaxation in the high-stress region increases the loop width.

Another kind of distorted loop shape is shown in Fig. 124. These

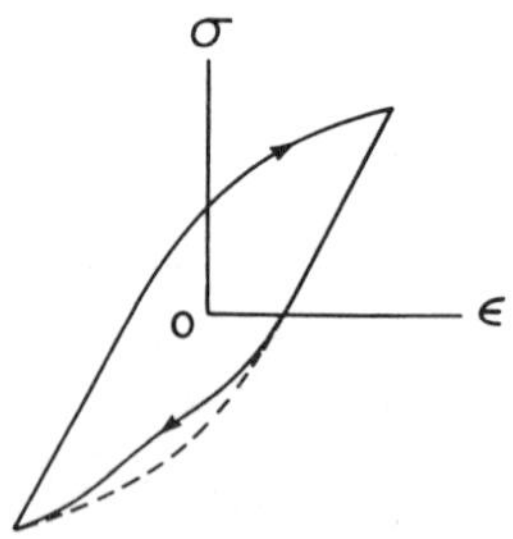

Fig. 124. Bendback in hysteresis loop of cast iron

are observed in cast irons [62, 78]. The tension part of the loop is quite normal, but going into compression there is a so-called bendback. It appears as if the metal wanted to change in midcourse from one hysteresis loop to another. The behavior is indeed a result of two different stress-strain responses. The dispersed graphite flakes make the metal weaker in tension than in compression. Such behavior may have a certain universal significance for any composite material that is composed of widely different materials.

8.2 Cyclic Creep without a Mean Stress

The general observation is that the necessary conditions for cyclic creep are load control, cyclic plasticity, and the presence of a mean stress. The counterpart of this behavior under strain control is cycle-dependent relaxation, which tends to reduce the mean stress toward zero stress. These expected behaviors are somewhat altered when the material has unequal strengths in tension and compression, as in Fig. 124. For example, cyclic creep can occur even when the imposed loading condition is such that the material is apparently not under a mean stress. Figure 125 shows the behavior of high-

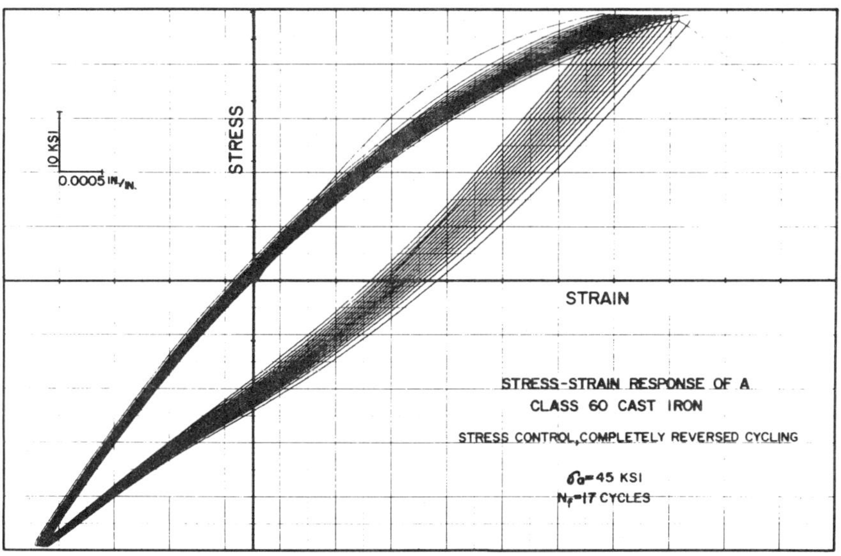

Fig. 125. Cyclic creep of cast iron under fully reversed stresses (from Richard and Sandor, "The Cyclic Stress-Strain Response of High Strength Cast Iron")

strength cast iron under a fully reversed stress control condition. There is cyclic creep as if the metal were under a tensile mean stress. It is believed that in this case the increasing mean strain is caused by the relative weakness in tension. In other words, there is always a little more plastic flow in the tensile direction than in compression, in the same way that there is under a tensile mean stress. It could be said that under fully reversed imposed loads this metal provides its own effective mean stress.

Of course, an argument could be made against this concept on the basis of a previously presented observation. In that it was shown that an extremely small mean stress can cause cyclic creep if the cyclic plasticity is large. Naturally, this is a possible factor in the creep shown in Fig. 125 but not necessarily a significant one, since the plastic strains are not too large. The possibility of a self-induced effective mean stress still exists, and it is supported by the following observations.

Under strain control a mean stress of either sign tends to relax toward zero stress. The larger the strain range, the smaller is the stable value of the mean stress. This behavior is shown schematically by the two broken curves in Fig. 126. In the case of the cast

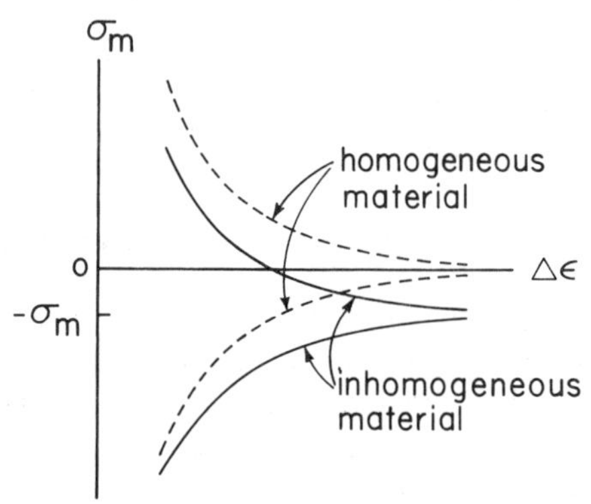

Fig. 126. Idealized curves for the relaxation of mean stresses

iron discussed above, a large cyclic strain control condition also produces a relaxation of an initially imposed mean stress. This mean stress is not relaxing toward zero stress, however, but rather toward a compressive stress of about 10 ksi [62]. This tendency is shown by the two solid curves in Fig. 126. In reality, fracture often occurs before the relaxation of the tensile mean stress is able to stabilize. Such behavior is just another evidence of a material's preferential strength, and should be further investigated with respect to the fatigue properties of composite materials.

8.3 Hardening and Softening under Constant-Amplitude Cycling

Cycle-dependent softening or hardening are common material instabilities. A given metal may do either one under constant-amplitude loading, depending on its initial condition. Both behaviors can occur in a given specimen when the loading amplitude is suddenly changed to a different level. It happens, though rarely, that a given specimen shows both extremes of cyclic adjustment under cycling at a constant amplitude of total strain. One of the possibilities is shown schematically in Fig. 127. The constant strain amplitude is the controlled quantity, and the stress amplitude varies independently. In the first few cycles there is hardening, followed by softening toward a stable level for the rest of the life. The initial hardening may be slight,

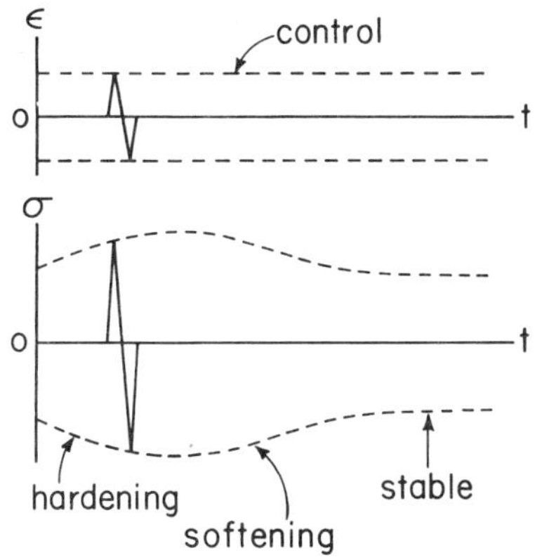

Fig. 127. Complex instability in stress-strain response

as in the case of a Ti-8Al-1Mo-1V alloy [14]. The reverse situation may also be observed where a relatively large amount of initial softening is followed by considerable hardening [29].

In either case, the behavior is caused by metallurgical instabilities that are not always clearly understood. It is probable that the competing processes of hardening and softening are always present. The observed macroscopic behavior is the net result of the two processes. Of course, the important aspect from the fatigue point of view is this net cycle-dependent variation in the stress-strain response.

8.4 Strengthening by Strain-Aging

It is generally believed that the strength of materials is inversely related to temperature, as shown by the solid curve in Fig. 128.

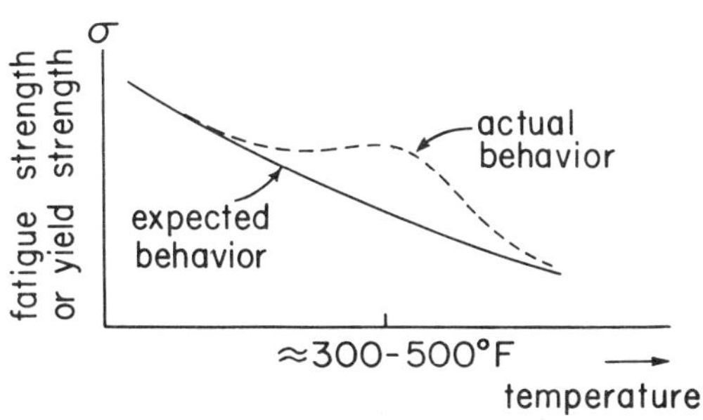

Fig. 128. Strengthening by strain-aging in low-carbon steels

There are many important exceptions to this pattern, especially in structural failures, where a low temperature may decrease the nominal fracture stress by causing embrittlement [72]. An interesting aspect of temperature-dependent behavior is the strengthening observed in low-carbon steels at moderately elevated temperatures. In such metals both the yield and fatigue strengths have a peak in the region of 300–500°F, as shown by the broken line in Fig. 128. The increase in strength can be explained by strain-aging, the interaction between dislocations and solute atoms of carbon and nitrogen [9, 11].

The resultant strengthening in fatigue depends somewhat on the cyclic-control condition. Since strain-aging increases the yield strength, the cyclic deformation resistance is increased. Therefore, under any controlled cyclic load, the plastic strain per cycle is decreased, and this increases the life. The improved deformation resistance is less useful under strain control, which relentlessly pushes and pulls the metal regardless of the internal resisting forces that may develop. Any improvement in fatigue life caused by cyclic strain-aging can be adequately explained on the basis of the reduction in cyclic plasticity [65].

8.5 Life Reduction by Electropolishing

Notches of all sizes are generally detrimental in fatigue. In brittle materials, even scratches of microscopic dimensions can significantly reduce the life. A sensible solution to this kind of problem is to strive to attain as smooth a surface finish as possible. It is surprising that surface-smoothing apparently can be overdone. For instance, a mirror finish produced by electropolishing generally reduces the fatigue life in metals.

An explanation of this behavior can be made on the basis of residual stresses. The most common finishing operations are mechanical, such as milling, grinding, or polishing with abrasives. Each one of these operations leaves numerous scratches on the surface. The depth of the flaws is a factor in the fatigue performance, but it is even more important that the mechanical polishing introduces a compressive residual stress in the near-surface layers of the material. An electropolishing treatment can reduce the life, because it not only removes the harmful scratches but also the beneficial compressively stressed layer. At the same time, the electropolishing does not create any stresses of its own. These observations show how important a role residual stress plays in fatigue.

8.6 Size Effect in Fatigue

It has been observed that under a given cyclic stress or strain control condition the fatigue life of a material is partially dependent on its volume. Experience shows that the larger the volume of material the shorter the life. Similar behavior occurs in monotonic loading to fracture, as Leonardo da Vinci observed long ago when he found that the fracture strength of iron wires decreased as the specimen length was increased, even though the cross-sectional areas were the same.

The simplest explanation of this phenomenon is that there is a greater chance for sizeable flaws to exist in the larger of the specimens. A complete treatment of the problem is difficult, because the notch and flaw dimensions, the stress gradients, and the size of the highest-stressed volume must be considered [19]. Of course, the importance of any contributing factor should be in its ability to enhance the cyclic plasticity, at least within small regions of the material. Direct relationships based on plastic strains are not yet available because the art of localized strain measurement is still inadequately developed.

8.7 Crack Growth in Compression

The beneficial nature of compressive stresses is generally recognized, but it can be easily overestimated. An interesting situation may develop, for instance, when a simple notched plate is axially loaded entirely in the compressive region, as shown in Fig. 129.

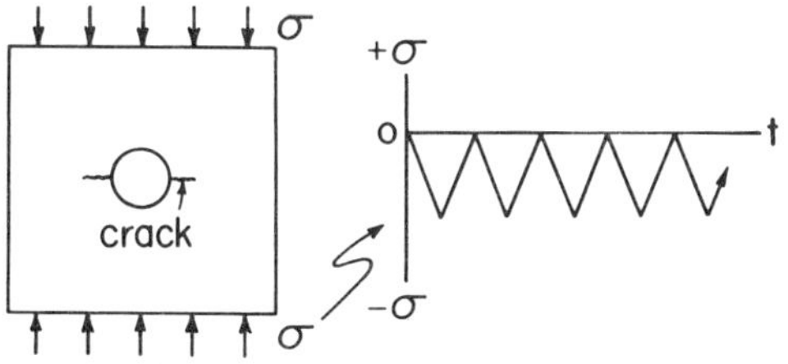

Fig. 129. Fatigue crack growth under compressive loading

Since compression tends to close and thus inactivate cracks, it is surprising to find that cracks can initiate at the hole and grow under the given cyclic loading. These cracks may become quite long compared to the plate width, but their growth is decelerating and may stop after many cycles of loading before complete separation occurs [30].

This behavior can be explained on the basis of stress concentration and the associated possibility of generating residual stresses. Every peak load in compression can be considered an overload producing plastic flow at the notch roots. Each subsequent unloading or near-zero load minimum creates a momentary tensile stress at the hole and later at the tips of the cracks. The behavior demonstrates dramatically that an overload can produce a residual stress of the opposite sign and of substantial magnitude. In the present example, the tensile residual stresses are generated without any external tension and yet they are large enough to cause cracking. The plastic strain per cycle may be quite small even at the notch root, but it is sufficient to reestablish some tensile residual stress in every cycle and allow damage accumulation to culminate in fracture. The analysis of this kind of behavior is also hampered by a lack of accurate cyclic strain measurements on the necessarily small scale of the critical zones.

8.8 Simultaneous Creep and Cyclic Deformation

Cyclic creep as it has been discussed here is by no means the only way in which a material can become unstable under the combined action of simultaneous stresses. It has been observed, for instance, that a metal can be locally indented under an extremely small static pressure when cyclic plastic deformations are imposed in a direction perpendicular to the indentation pressure [7]. Another similarly interesting observation is that the axial creep rate in a cylindrical member may be increased by several orders of magnitude when torsional oscillations are also imposed on the member [21]. Whatever the circumstances of cyclic creep, it is clear that there is a mutual enhancement of static and cyclic deformations. Either one or both of these applied loading conditions may be in tension (or compression) or in shear. One could say that inelastic deformation caused by a force acting in a given direction increases the material's ability to deform inelastically under loading in other directions. The fundamental description of the damage accumulation process on the basis of cyclic plastic strain should not be affected by the state of stress that produces the particular plastic strains. In the final analysis, an increase in plasticity can be directly related to a higher rate of fatigue damage accumulation.

8.9 Fatigue of Electroplated Parts

It is frequently found that a thin layer of electrodeposited metal reduces the fatigue life of the original component. This is surprising for the following several reasons. The volume of the electroplate is relatively small, so that an easily noticeable change in life caused by this material is not expected. Metals commonly used for plating are quite effective in protecting the base metal from corrosive attack, and this is definitely a beneficial factor which should increase the fatigue life. Nevertheless in some cases the life is reduced, even though the plating is the strongest of the two metals and therefore strengthening might be expected from the metal deposit. Each one of these considerations is partly valid, but there is still a competition between beneficial and harmful effects that determines the fatigue behavior.

When the harmful processes are obviously dominant (because the life is reduced), the cause is found in the stress-strain behavior of the individual metals and in the response of their combination to external loads. The general observation has been that the life is decreased by a relatively hard electroplate such as chromium on steel. When cyclic loads are controlled on the plated member, it is the bulk of the member that must carry the load, regardless of the strength of the plating. One could say that the large mass of the base metal ignores the deformation resistance of the thin plating. Thus, the plated metal is essentially under strain control whose magnitude is determined by the base metal's response to the external load. This is very much like the localized control conditions in notched members.

The hard plating is less than ideal under strain control because it cannot deform plastically to reach the strain limits without developing high levels of stress. The high stress and low tolerance to plastic flow make it likely that cracking will begin within the plating metal, probably at its free surface. When these cracks reach the interface, the substrate metal is weakened by the stress concentrations generated at its boundary. The overall fatigue life is reduced because in the hard plating the cracks initiate easily and, once started, propagate by stress concentration into the interior of the member. Of course it would be incorrect to generalize from this behavior and claim that electroplating always decreases the fatigue resistance.

8.10 Fatigue of Atomic-Chain Fiber

As a last example, consider an imaginary test with a specimen of atomic proportions. The specimen may be called the ultimate in thin fibers, since it is a single chain of identical atoms, shown schematically in Fig. 130. The diameter of the fiber is the diameter

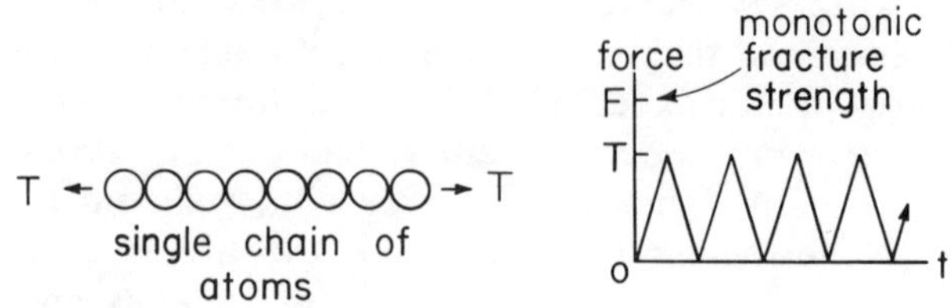

Fig. 130. Cyclic loading of hypothetical atomic-chain specimen

of an atom. Let us assume that this fiber is axially loaded by a pulsating force that varies between zero and a tension T as shown in the diagram. Assume also that in monotonic tension the filament breaks when the force reaches the level F. Three different cyclic tests under constant amplitude loading are visualized with $T_1 = 0.01F$, $T_2 = 0.5F$, and $T_3 = 0.99F$. Let us consider, qualitatively speaking, what the difference is in specimen response under these three loading conditions.

The applied loads are cyclic, but one should not expect a fatigue failure in any arbitrary number of cycles under either one of the given cyclic forces. The reason for this is that there is no way for damage of any kind to accumulate from cycle to cycle in this extremely simple specimen. The atomic forces that hold the chain together always act in the same manner regardless of prior repetitions of loads, as long as these loads are less than the monotonic fracture strength F. Therefore the chain can be broken only by a force F. The life is infinity under any cyclic loads smaller than F, because the atoms can always rebound to their positions of equilibrium when the load is removed.

This hypothetical problem illustrates an important aspect of real fatigue situations. A fatigue problem can exist only when some physical quantity accumulates as cycling progresses. This cumulative quantity, of course, is plastic strain.

9

The Approaches to Fatigue Problems

The best practical approach to a specific fatigue problem depends not only on one's knowledge of the subject but also on a number of other considerations. These considerations are based on the degree of reliability required and the financial and time limitations imposed. Often several different approaches are taken simultaneously. The following discussion is divided into sections for clarity and to indicate possible differences in the degree of human control over the fatigue damage process. There is no intention in making these groupings to judge the relative merits of the various methods. In general, the greatest benefits can be expected when the human control is extensive. This, however, requires a large investment of engineering resources.

9.1 Post-mortem Examinations

In the early part of the history of fatigue knowledge (before *S-N* curves), the most significant information was obtained simply by looking at the final fracture surfaces. The initial concept derived from such observations was that the metal crystallized during cyclic loading, causing failure at loads below the ultimate strength. This is, of course, an erroneous thought, since all metals are crystals in the first place. In spite of the original misconception, fractography has been (and will remain) a useful technique in fatigue analysis.

Much information about the failure process can be revealed by the macroscopic and microscopic features of the fracture surface [28, 33, 45, 52, 83]. The two factors of greatest interest are the origin of the fatigue crack and its size with respect to the whole fracture area. The fatigue crack can be distinguished because it is relatively flat and essentially perpendicular to the direction of the imposed tensile stress. These features are the same as those of brittle cleavage fractures in monotonic loading. Therefore, the fatigue fracture process seems to be a brittle one, in apparent contradiction to the plastic deformation whose importance has been emphasized throughout this text.

A simple explanation of the problem can be made by considering the difference between cracking and the damage accumulation prior to the fracture. A crack is damage, of course, but it is only the final manifestation of a less dramatic cumulative process. This process is the cyclic plasticity that is responsible for the exhaustion of ductility, at least in small localized regions of the material. Once the available ductility is used up, the material can absorb energy only by creating new surface area [72]. In this way even a ductile metal can be imagined to fracture in a brittle manner. The real behavior may be more complex, however. The macroscopically brittle fracture appearance does not preclude the possibility of numerous microscopic shear failures, which have actually been observed. In either case, plastic deformations are prominent in the damage accumulation process.

On closer examination the fatigue crack may reveal the point of its origin. The crack nucleus often appears to be the source of wavelike ripples on the crack surface, as shown in Fig. 131.

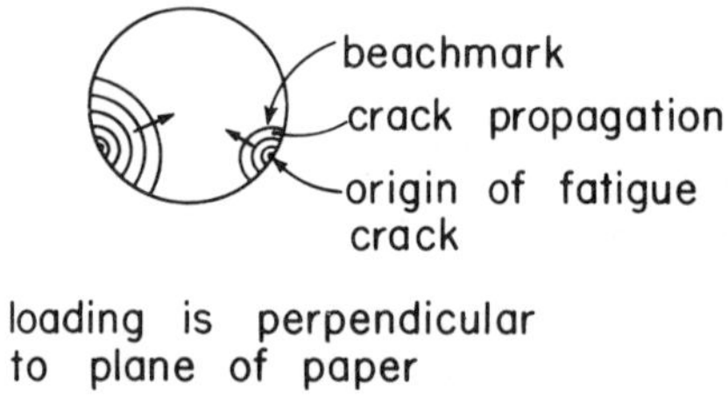

Fig. 131. Fatigue fracture surface appearance; note that several cracks may start simultaneously

There may be several original cracks at a cross section. The ripples are called "beachmarks" and can often be distinguished by the unaided eye. In some cases there are also fine ripple features that can be seen only with the aid of the electron microscope. These fine markings represent stepwise advances of the crack front. Occasionally it is distinguishable that each ripple corresponds to one cycle of the load. Figure 132 shows such an advance of the crack front in multiple-level loading.

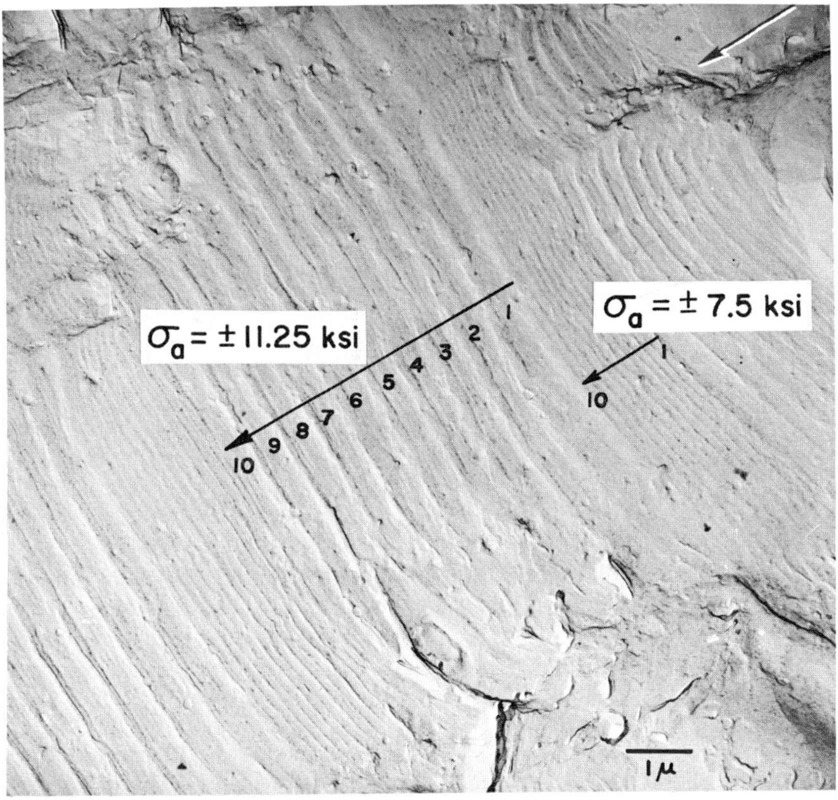

Fig. 132. Characteristics of spectrum loading. Mean stress = 13.75 ksi; three alternating stresses (2.5, 7.5, and 11.25 ksi) applied in 10-cycle intervals. Note lack of striations for lowest alternating stress (2.5 ksi). Arrow indicates fracture direction (× 13,800; from Whiteson, Phillips, Kerlins, and Rawe, in *Electron Fractography*. Courtesy ASTM)

Beachmarks have immediate practical value in failure investigations. They show the point of crack initiation at an inclusion, flaw, or any other stress concentration. Design information obtained in this manner and the corrections made are costly and time-consuming but frequently unavoidable. The simplest solution to a fatigue problem discovered this way is to add more material at the critical section in order to reduce the operating stresses. A more satisfactory method is to direct attention first of all to the stress concentration where the failure initiated. It may be found that a reduction of the stress concentration, even at the price of losing some of the load-carrying material, can extend the fatigue life, as shown in Fig. 133.

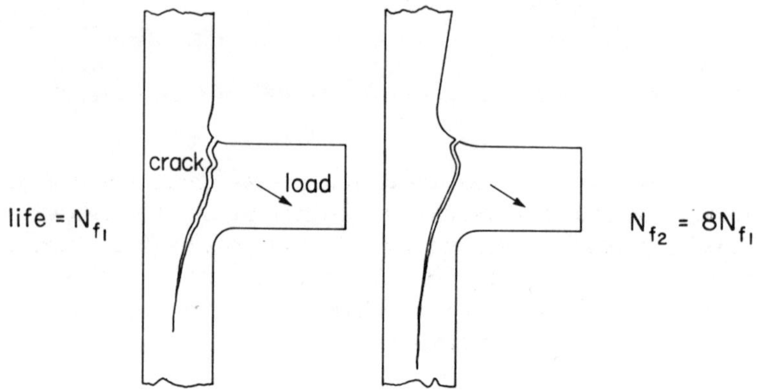

Fig. 133. Improvement of fatigue life by reduction of stress concentration (after Smith, *Tips on Fatigue*)

The simple approach of adding more material is often practical because in most engineering problems the imposed forces are the controlling factors. Thus more material means lower stresses and longer life. It must be emphasized, however, that the benefit of having more material is strictly dependent on the applied control condition. Two possible situations will be discussed to illustrate the problem. Figure 134*a* shows a cantilever beam subjected to

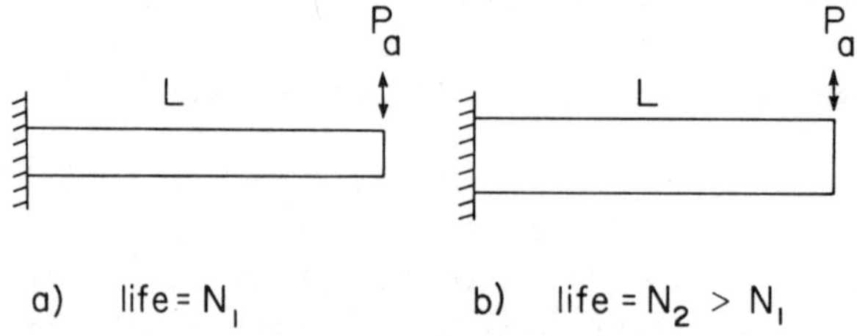

Fig. 134. The effect of beam thickness on life under cyclic load control

cyclic forces of a constant amplitude P_a. A deeper beam of the same material and same length (Fig. 134*b*) has a longer fatigue life under the same amplitude P_a. The life N_2 is longer than N_1 because the added material in the deeper beam has reduced the stress throughout. Under the deflection control shown in Fig. 135, the thinner of the two beams has the longer fatigue life. This is

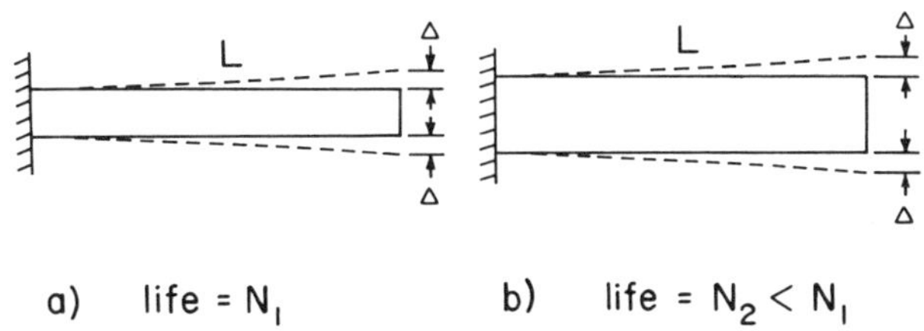

Fig. 135. The effect of beam thickness on life under deflection control; note that Δ in *a* = Δ in *b*

true because, for a given deflection Δ, the peak stresses (at the specimen surface in bending) are largest in the thick beam. Thus the cyclic plastic strains are also largest in the thick beam. This indicates why full-scale testing will remain a useful method in fatigue investigations until complex problems of control condition and the material response can be analyzed in detail.

9.2 Fatigue-Damage Prediction and Indication

A high degree of control over the fatigue problem is achieved if the time and place of failure can be predicted accurately [46]. This is partially a design problem, but generally design planning must be complemented by a monitoring of some cycle-dependent change in the material. The most obvious indication of damage is the fatigue crack. Methods for systematically searching these out can range from crude visual observations through the more satisfactory magnetic, eddy current, ultrasonic, and X-ray testing. The literature on nondestructive testing is rich in descriptions of various methods. It should be emphasized that an easily detectable fatigue crack often appears only in the last 5 or 10 percent of the life. Thus sufficiently early detection of cracks is always difficult and expensive.

Cracking is, of course, only the final manifestation of the damage accumulation from plastic strain that goes on from cycle to cycle before fracture occurs. An intriguing method of measuring the cumulative effect of cyclic loading is based on the use of fatigue life gages [27]. These are like strain gages in physical appearance but respond to deformation by an irreversible change in electrical resistance. A small gage of this kind can be attached to each known and approachable critical region of the structure. Simple resistance measurements at any time during the service life indicate by the magnitude of the resistance the damage summation up to that time, but these measurements are not yet very reliable. Among other things, the important factors affecting the measurements are the material properties of the structure, its geometry, environment, and the life range of operation. Further research in this area may make the life gages more universally applicable than they are at the present.

The fatigue life gage belongs in the general category of mechanical fuses, which includes any small piece of material that is riding along with the deformations of a larger structural member. The fuse is not a critical component itself, so it can be designed to fail before the large member does. This failure signals that loading of the system has become unsafe.

The fundamental approach to life prediction should be based on an analysis of cyclic plastic strain as it has been discussed under cumulative damage. The essence of this method is to know the damaging potential of any level of plastic strain, and to know the cycle-dependent function of local plastic strain at the most critical region. For single-level or periodically repeated blocks of loading, the required function may be established by measurements taken early in the life. Unfortunately, the available instrumentation is still often inadequate to provide the necessarily precise data for using this fundamental method.

9.3 Material Selection

Perhaps as important as the ability to predict the fatigue life is the ability to select the material that will give the longest life under the given circumstances. From the point of view of mechanics, the problem is to match the material with the expected loading-control condition for optimum results. The analysis of the control condition should take into account the possibility that the externally imposed condition is not necessarily the local condition at every point in the material. The selection of an appropriate material can be made only if a great deal is known about every proposed material and the required service condition. In some designs this is not critical; several different materials may be chosen with satisfactory results. It cannot be overemphasized, however, that truly efficient use of a material is possible only when many of its mechanical properties are well known.

The preceding chapters have described several important properties that should be considered in fatigue. The material-characterization sheets that follow show a simple way of systematically presenting and storing information about these properties. All of the entries on the sheets may not be needed for every design problem, but it is also possible that the list may occasionally be incomplete. Acquiring the information shown is a substantial task when many

MATERIAL CHARACTERIZATION SHEET 1

Materials: SAE 1045 Specimen: Longitudinal, 0.5" x 0.2"

Condition: Q&T (1200°F) - 225 BHN Test Condition: Room temp.

Monotonic Properties:

Mod. of Elast., E	29	$\times 10^3$ ksi
Sheer Mod., G		$\times 10^3$ ksi
Yield Strength, 0.2% Sy	92/110	ksi
Ultimate Strength, Su	105	ksi
Red. in Area, % RA	65	
True Frac. Strength, σ_f	209/178	ksi
True Frac. Ductility, ϵ_f	1.04	
Strain Hrd. Exp., n	0.13	
True Toughness, U_p	165,645	in-lb/in^3

Comments: Values are averages for six specimens.

Cyclic Properties:

Yield Strength, S_y'	60	ksi
Strain Hrd. Exp., n'	0.18	
Strength Coeff, K'	195	ksi
Fatigue Strength Coeff, σ_f'	178	ksi
Fatigue Ductility Coeff, ϵ_f'	1.0	
Fatigue Strength Exp, b	-0.095	
Fatigue Ductility Exp, c	-0.66	
Transition Fat. Life, $2N_t$	10,000	rev

Comments: 1. closed-loop, servo control
2. three specimens for each life range: 50, 500, 10^3, 10^4, 10^5, 10^6, 10^7 reversals.

Source: R.L., B.I. Date: 1/70

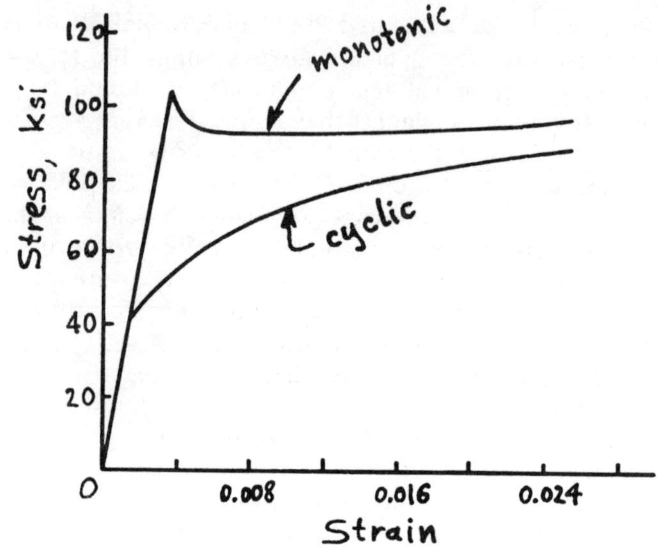

Stress, ksi
120
100
80
60
40
20
0
0.008
0.016
0.024
Strain
monotonic
cyclic

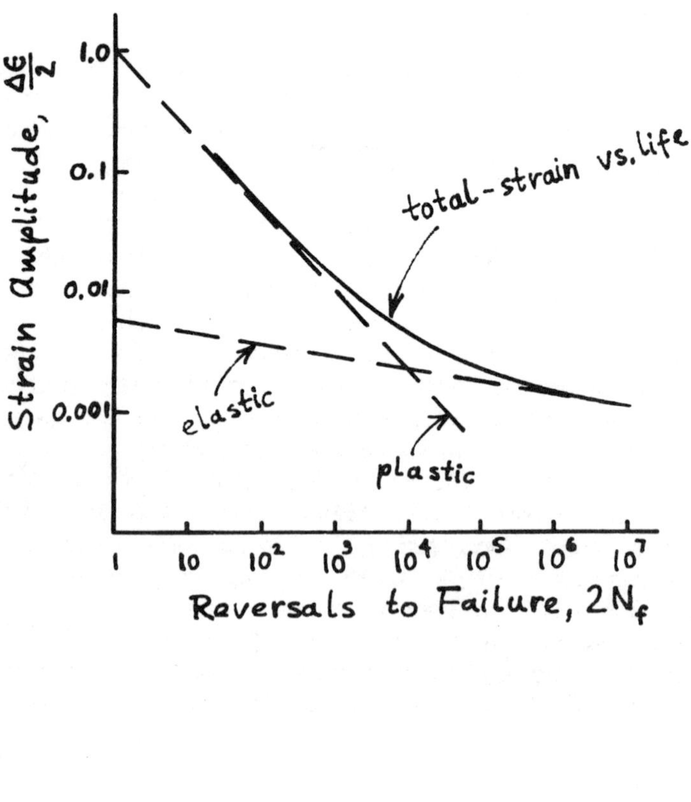

Strain Amplitude, Δε/2
1.0
0.1
0.01
0.001
1
10
10²
10³
10⁴
10⁵
10⁶
10⁷
Reversals to Failure, 2Nf
total-strain vs. life
elastic
plastic

materials have to be evaluated, but such an initial investment is worthwhile because the data are always applicable to the material tested under the specified test conditions. It should be invaluable to be able to tell at a glance that a metal severely softens when cyclically loaded, as in the example shown. The value of the simple diagrams and numerical factors is, of course, enhanced by the concise presentation of any relevant comments that add to the information on the material. For example, it could be recorded that a relatively small difference in operating temperature may cause a significant change in the strain-hardening behavior of a metal that is somewhat similar to the one in question.

In selecting a material for a particular service, it must be remembered that a ductile material is best to resist (actually, to tolerate) plastic deformations without failure, while a strong material behaves quite elastically under a load. A so-called tough material is ideal because it can deform plastically while sustaining high stresses. The relative fatigue behaviors of ductile, strong, and tough materials are shown schematically in Fig. 136. The sketches show that the life range may require a significant consideration. At long lives, for example, the plastic strains are very small and therefore the ability of the material to tolerate plastic flow is less important than its strength. Of course, failure is still caused by plastic strains.

The curves for most metals cross each other at about 1000 cycles of life when the strain amplitude is 0.01 [17, 34]. This approximately common point on the life plot (Fig. 136*a*) cannot be explained on a quantitative physical basis. Qualitatively, it shows that high strength and ductility are to some degree mutually exclusive in a single material. Such a conclusion appears logical, but it is unnecessarily pessimistic. There is much evidence that it is possible to obtain high strength and good ductility in steel. The indications are that further improvements in this area can be expected [31, 32, 84].

The matter of strength and ductility is extremely important in any situation where fracture could occur. From the practical point of view, it is not absolutely necessary to have the ideal of high strength and ductility simultaneously and everywhere in the material. The reason is that most engineering structures are primarily load-transmitting devices in which high ductility is required only at small regions of stress concentration. For further discussion of the problem let us consider a simple model such as the plate with a circular hole in Fig. 137. The imposed stress σ may be static or cyclically changing. Regardless of the virtues of ductile materials, the plate has to resist the applied load without failure. This can be accomplished only if a large portion of the minimum cross section is strong.

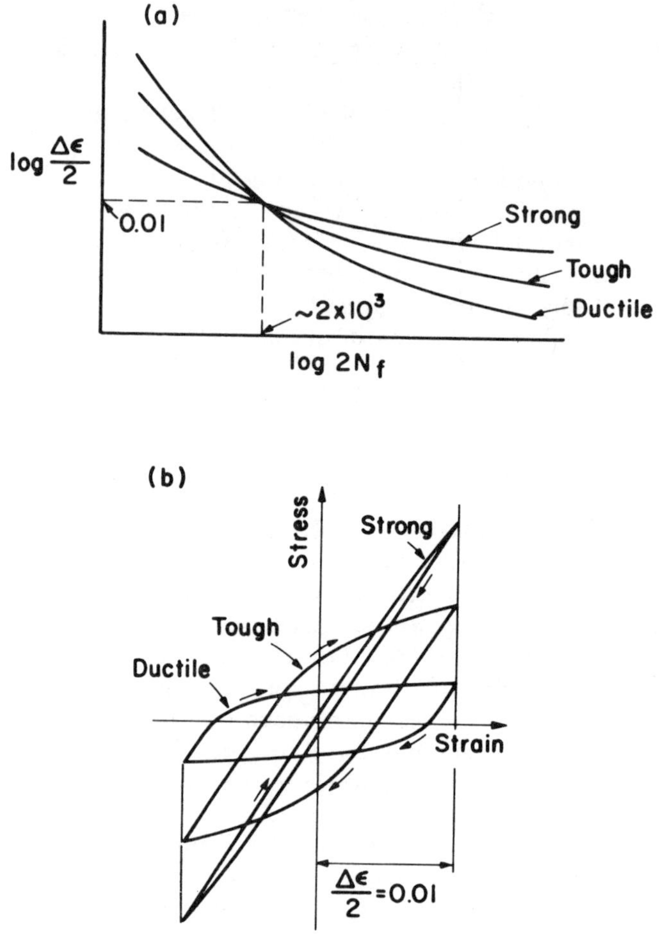

Fig. 136. Schematic representation of cyclic strain resistances (from Landgraf, *High Fatigue Resistance in Metals and Alloys*. Courtesy ASTM)

The stress concentration of the hole makes it desirable to allow plastic deformation in the critical regions. Under cyclic external load control the notch region experiences strain control, which also makes it desirable to have ductility at the hole.

Thus the ideal situation for a complex member is to have the strong and ductile parts segregated from each other, as shown in Fig. 137. In this arrangement, the two different regions help each other as follows. The strong part carries most of the load without much deformation. The stress concentration is blunted by plastic flow, so that crack initiation and possible catastrophic failure of

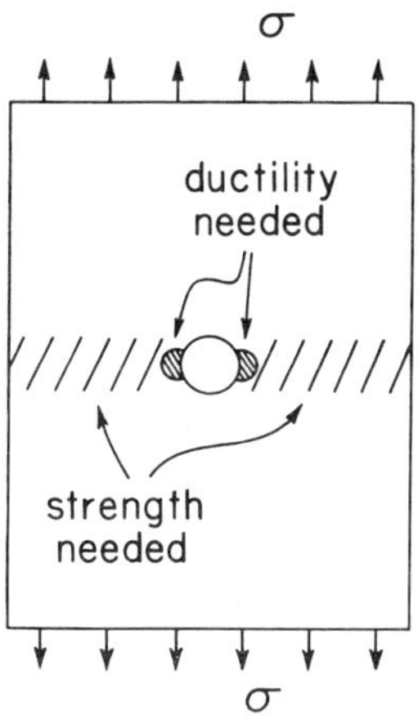

Fig. 137. The desired segregation of strength and ductility at the minimum section

the whole piece is prevented. At the same time, the plastic strains in the critical region are not excessive because the surrounding strong material acts as a constraint to deformation. Therefore high strength in the notch region is not required, although coupled with high ductility it would raise the fracture toughness [72].

The artificial distribution of flow strengths within a given structural member requires the use of a composite material. The composite may be of two different materials. A single material may also serve, if its mechanical properties can be appropriately varied throughout the member. Varying the grain size offers a chance to obtain the desired properties, for example. Fatigue strength (under load control) depends on grain size, as shown in Fig. 138. Thus the overall load-carrying ability of a complex member can be increased if the grain size is small everywhere except in the critical regions of stress concentration.

Of course this brief discussion cannot cover all of the many problems associated with material selection. The aim here is to emphasize a few basic concepts. One is that high strength is desirable when loads must be transmitted, while a ductile material is best suited to endure imposed deformations without failure. Another point worth remembering is that matching the material with the applied control condition must be done by taking into account the situations existing in small regions of critical behavior.

9.4 Fatigue-Damage Alleviation

The highest possible level of human control over the fatigue problem would be achieved if fatigue failures could be delayed indefinitely

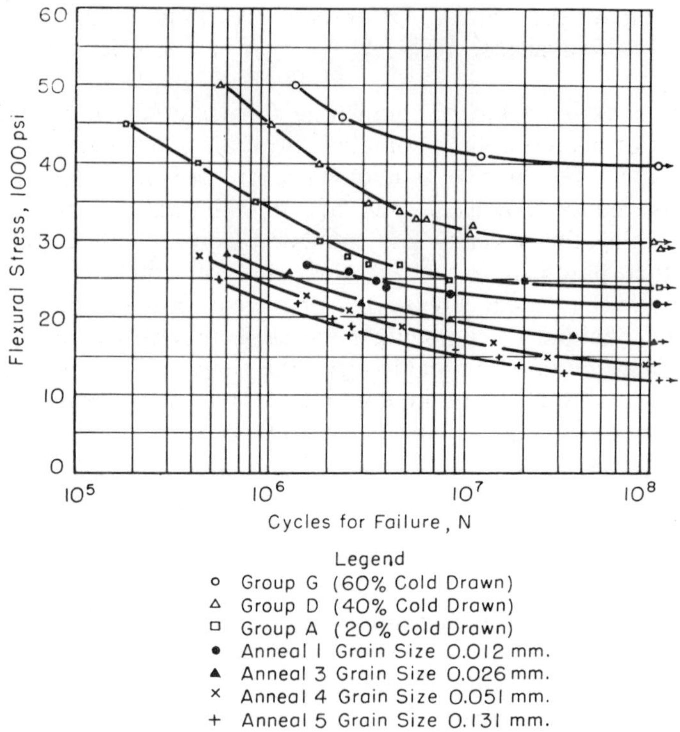

Fig. 138. Influence of grain size and cold work on the flexural fatigue strength of 70–30 brass (from Sinclair and Craig, "Influence of Grain Size on Work Hardening and Fatigue Characteristics of Alpha Brass." Courtesy ASM)

[46, 66]. Ideally, this would not mean that all operating stresses should be low enough to allow infinite life. For economical use of a material, the stresses imposed on it should be as high as possible, which means that damage accumulation is unavoidable. The life can be extended under this condition only if the damage is periodically repaired or if the rate of damage processes is artificially reduced [66].

Repairing the damage by healing the cracks is one possible approach. For example, the crack surfaces may be pressed together for cold welding. This is not very practical as an artificial repairing method because oxidation of the surface cracks prevents healing in this critical part of the material. The rough crack faces also prevent adequate matching of the surface features when they are pressed together.

Another possible but essentially unexplored method of crack healing is by filling the crevices with a material that can retard the propagation of the cracks. Obviously, this kind of process could affect only the available surfaces, but that is where cracks normally initiate.

The most extreme way of eliminating surface cracks is by removing the layer of material that contains the cracks or that is ready to fail [60, 73]. This method is practical, even though the load-carrying matter is reduced, because the life of the material is improved by decreasing the severity and number of stress concentrations in it. Another practical aspect of this kind of crack elimination is that beneficial residual stresses can be created in the process. The simplest technique of removing material is mechanical polishing or other remachining, which cold-works the surface and leaves it compressively stressed. Even in low-cycle fatigue, the remachining is effective in extending the fatigue life [60]. One practice of remachining occasionally used is illustrated in Fig. 139. A hole can be redrilled

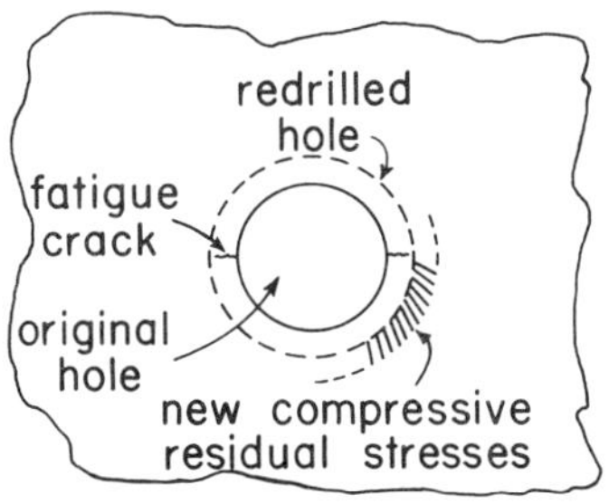

Fig. 139. Fatigue damage alleviation by remachining

to a slightly larger size when cracks are observed. This blunts the stress concentrations and leaves a compressively stressed layer all around the hole. Such a method is valuable if the cracks are not allowed to grow long. The redrilling may have to be performed quite frequently for optimum results. Unfortunately, there is not yet enough information available to establish predictable results from these intermittent repair procedures.

The beneficial compressive residual stresses obtained by mechanical remachining can be periodically recreated, of course, without waiting for cracks to appear. Intermittent peening operations can reintroduce the desirable stresses many times during the fatigue life if these stresses tend to fade out because of cyclic plastic strains. Naturally, this is damage prevention rather than actual repair. Such a treatment is reasonable only if the plastic strains are not large. Otherwise, the residual stresses can be rapidly eliminated.

It is obvious that the cause of any delay in the accumulation of fatigue damage is the controlled level of cyclic plasticity. Some quantitative examples of extending the fatigue life by reducing the plastic strains are obtained in cyclic strain-aging of low-carbon steel tested under stress control [65]. Strain-aging strengthens the metal so that under a given stress amplitude the cyclic plasticity is decreased. The resulting extension in life is accurately predictable by using the cumulative damage calculations based on the actual plastic strains and the fatigue ductility exponent. Such examples strengthen the belief that cyclic stress is the cause of fatigue damage only to the extent that it causes plastic strains.

The role of the fatigue ductility exponent in artificially influencing the fatigue life is worth remembering. For most metals the magnitude of the exponent is approximately 0.5. Thus a rule of thumb may be stated relating the change in plastic strains to the expected life. For a given metal, the factor of the increase in life is about the square of the factor of the decrease in plasticity, regardless of the stress amplitude. For example, if one wants to increase the life by two orders of magnitude, this requires an approximate reduction in cyclic plastic strain amplitude by one order of magnitude. Naturally this is difficult to accomplish.

None of the possible damage alleviation methods is a final solution to fatigue problems in general. Nevertheless these methods are valuable for two reasons. First, they can help in developing a certain new understanding of the causes of fatigue damage. The results of intermittent surface treatments during the life show the relative significance of damage initiation at the surface, for instance. Second, damage alleviation offers a chance to steadily improve the fatigue performance of materials. Whether a method is practical or not depends on many factors that are obvious to the designer. The potential of the approach may be illuminated by the crude analogy of the refueling concept. Moving vehicles seldom carry all the fuel necessary during their total lifetimes. It may be envisioned that certain load-carrying structures could also be periodically restored with load-carrying ability. This would contrast with the current practice of providing at the outset all the strength or ductility needed by structural and machine members for their entire lives.

Conclusions

There are two ideas that are immensely useful in dealing with material fatigue. One is that the stress-strain response of a material can be altered by cyclic loading. The fatigue life, crucial as it may be, is only the final expression of a material's cyclic behavior. The other concept is that fatigue damage is caused by cyclic plasticity. Cyclic stresses cause damage only as much as they plastically deform the material.

There is enough information available at the present to help one appreciate the role of plastic strains in fatigue. Yet the current technology is not adequate for the accurate measurement of cyclic plastic strains in a sufficient variety of situations. Two important problems, in particular, may be mentioned. It is not yet possible to measure and control very small cyclic plastic strains routinely. Difficulties generally arise when these strain amplitudes are about 10^{-4} or less. Another serious problem occurs even with large plastic strains if they are concentrated in very small volumes such as in the notch root regions. Naturally, the challenge in instrumentation is greatest when extremely small strains have to be measured in highly localized regions. Precise analysis and prediction of the fatigue damage process will be possible only when satisfactory solutions become available to these difficult problems.

REFERENCES

INDEX

References

1. Almen, J. O., and Black, P. H., *Residual Stresses and Fatigue in Metals*, McGraw-Hill, New York, 1963.
2. Avery, D. H., and Backofen, W. A., in *Fracture of Solids*, John Wiley & Sons, New York, 1963.
3. Benham, P. P., "Fatigue of Metals Caused by a Relatively Few Cycles of High Load or Strain Amplitude," *Metallurgical Reviews*, Vol. 3, No. 11, 1958, pp. 203–234.
4. Byars, E. F., and Snyder, R. D., *Engineering Mechanics of Deformable Bodies*, International Textbook Co., 1969.
5. Coffin, L. F., Jr., "Low Cycle Fatigue: A Review," *Applied Materials Research*, Vol. 1, No. 3, Oct. 1962, pp. 129–141.
6. Coffin, L. F., Jr., "The Influence of Mean Stress on the Mechanical Hysteresis Loop Shift of 1100 Aluminum," *Journal of Basic Engineering* (Series D of *Transactions of the ASME*), Vol. 86, Dec. 1964, pp. 673–680.
7. Coffin, L. F., Jr., "The Stability of Metals Under Cyclic Plastic Strain," *Journal of Basic Engineering* (Series D of *Transactions of the ASME*), Vol. 82, Dec. 1960, pp. 671–682.
8. Corten, H. T., and Dolan, T. J., "Cumulative Fatigue Damage," *Proceedings of the International Conference on Fatigue of Metals*, The Institution of Mechanical Engineers, London, 1956, pp. 235–246.
9. Cottrell, A. H., *Dislocations and Plastic Flow in Crystals*, Oxford University Press, Oxford, 1953.
10. Crews, J. H., Jr., and Hardrath, H. F., "A Study of Cyclic Plastic Stresses at a Notch Root," *Experimental Mechanics*, Vol. 6, No. 6, June 1966, pp. 313–320.
11. Dieter, G. E., Jr., *Mechanical Metallurgy*, McGraw-Hill, New York, 1961.
12. Dolan, T. J., "Designing Structures to Resist Low Cycle Fatigue," Paper No. W71-22.3, ASM-SME 1971 Western Metal and Tool Conference, Los Angeles, March 1971.
13. Dolan, T. J., "Nonlinear Response Under Cyclic Loading Conditions," *Proceedings of the Ninth Midwestern Mechanics Conference*, Madison, Wisconsin, Aug. 1965, pp. 2–21.

14. Endo, T., and Morrow, JoDean, "Cyclic Stress-Strain and Fatigue Behavior of Representative Aircraft Metals," *Journal of Materials,* Vol. 4, No. 1, March 1969, pp. 159–175.
15. Eul, W. A., and Woods, W. W., "Shear Strain Properties to 10^{-10} of Selected Optical Materials," NASA Report CR-1257, The Boeing Company for the National Aeronautics and Space Administration, 1969.
16. *Fatigue Design Handbook,* ed. J. A. Graham, Society of Automotive Engineers, New York, 1968.
17. Feltner, C. E., and Landgraf, R. W., "Selecting Materials to Resist Low Cycle Fatigue," ASME Paper No. 69-DE-59, American Society of Mechanical Engineers, 1969.
18. Feltner, C. E., and Morrow, JoDean, "Microplastic Strain Hysteresis Energy as a Criterion for Fatigue Fracture," *Journal of Basic Engineering* (Series D of *Transactions of the ASME*) Vol. 83, March 1961, pp. 15–22.
19. Forrest, P. G., *Fatigue of Metals,* Pergamon Press, Oxford, 1962.
20. Freudenthal, A. M., and Heller, R. A., "On Stress Interaction in Fatigue and a Cumulative Damage Rule," *Journal of Aero-Space Science,* Vol. 26, No. 7, 1959, pp. 431–442.
21. Gain, B. R., and Sinclair, G. M., "The Effect of Temperature on Cycle Dependent Deformation," ASME Paper No. 66-MET-7, American Society of Mechanical Engineers, 1966; See also Report No. 273, Department of Theoretical and Applied Mechanics, University of Illinois, Urbana, Dec. 1964.
22. Gough, H. J., *The Fatigue of Metals,* Scott, Greenwood, and Son, London, 1924.
23. Grosskreutz, J. C., in *Fatigue: An Interdisciplinary Approach,* Syracuse University Press, Syracuse, New York, 1964.
24. Grover, H. J., *Fatigue of Aircraft Structures,* NAVAIR 01-1A-13, U.S. Government Printing Office, Washington, D.C., 1966.
25. Halford, G. R., "The Energy Required for Fatigue," *Journal of Materials,* Vol. 1, No. 1, March 1966, pp. 3–18.
26. Hardrath, H. F., in *Fatigue: An Interdisciplinary Approach,* Syracuse University Press, Syracuse, New York, 1964.
27. Harting, D. R., "The —S/N— Fatigue-life Gage: A Direct Means of Measuring Cumulative Fatigue Damage," *Experimental Mechanics,* Vol. 6, No. 2, Feb. 1966, pp. 19A–24A.
28. Hertzberg, R. W., in *Fatigue Crack Propagation,* ASTM STP 415, American Society for Testing and Materials, 1967, pp. 205–225.

29. Hinton, R. W., "Stress Relaxation and Cyclic Hardening of 12Ni-5Cr-3Mo Maraging Steel During Low Cycle Fatigue," *Transactions Quarterly*, American Society for Metals, Vol. 61, 1968, pp. 176–183.
30. Hubbard, R. P., "Crack Growth Under Cyclic Compression," *Journal of Basic Engineering* (Series D of *Transactions of the ASME*), Vol. 91, Dec. 1969, pp. 625–631.
31. Ivanova, V. S., and Gordienko, L. K., *New Ways of Increasing the Strength of Metals*, The Iron and Steel Institute, England, 1968. Translation of 'Novye Puti Povyshenie Prochnosti Metallov,' originally published by Izdatelstvo 'Nauka,' Moscow, 1964.
32. Koppenaal, T. J., "The Current Status of Thermomechanical Treatment of Steel in the Soviet Union," *Transactions Quarterly*, American Society for Metals, Vol. 62, 1969, pp. 24–37.
33. Laird, C., in *Fatigue Crack Propagation*, ASTM STP 415, American Society for Testing and Materials, 1967, pp. 131–180.
34. Landgraf, R. W., in *High Fatigue Resistance in Metals and Alloys*, ASTM STP 467, American Society for Testing and Materials, 1970, pp. 3–36.
35. Landgraf, R. W., "Effect of Mean Stress on the Fatigue Behavior of a Hard Steel," Report No. 662, Department of Theoretical and Applied Mechanics, University of Illinois, Urbana, Jan. 1966.
36. Landgraf, R. W., Scientific Laboratory, Ford Motor Co., Dearborn, Mich. Private communication, 1970.
37. Landgraf, R. W., Morrow, JoDean, and Endo, T., "Determination of the Cyclic Stress-Strain Curve," *Journal of Materials*, Vol. 4, No. 1, March 1969, pp. 176–188.
38. Manson, S. S., "Fatigue: A Complex Subject—Some Simple Approximations," *Experimental Mechanics*, Vol. 5, No. 7, July 1965, pp. 193–226.
39. Manson, S. S., "Interfaces Between Fatigue, Creep, and Fracture," *International Journal of Fracture Mechanics*, Vol. 2, No. 1, 1966, pp. 327–363.
40. Manson, S. S., *Thermal Stress and Low-Cycle Fatigue*, McGraw-Hill, New York, 1966.
41. Manson, S. S., Freche, J. C., and Ensign, C. R., in *Fatigue Crack Propagation*, ASTM STP 415, American Society for Testing and Materials, 1967, pp. 384–412.
42. *Manual on Low Cycle Fatigue Testing*, ASTM STP 465, American Society for Testing and Materials, 1969.
43. Marin, J., Borachia, P., and Rimrott, U. A., "The Effect of

Stress Cycling on the Static Mechanical Properties of SAE 4340 Steel," *Proceedings,* American Society for Testing Materials, Vol. 59, 1959, pp. 662–673.

44. Martin, D. E., "An Energy Criterion for Low-Cycle Fatigue," *Journal of Basic Engineering* (Series D of *Transactions of the ASME*) Vol. 83, Dec. 1961, pp. 565–571.
45. McMillan, J. C., and Hertzberg, R. W., in *Electron Fractography,* ASTM STP 436, American Society for Testing and Materials, 1968, pp. 89–123.
46. *Metal Fatigue Damage: Mechanism, Detection, Avoidance, and Repair,* ed. S. S. Manson, ASTM STP 495, American Society for Testing and Materials, 1971.
47. Miner, M. A., "Cumulative Damage in Fatigue," *Journal of Applied Mechanics,* Vol. 12, 1945.
48. Morrow, JoDean, in *Internal Friction, Damping, and Cyclic Plasticity,* ASTM STP 378, American Society for Testing and Materials, 1965, pp. 45–87.
49. Morrow, JoDean, "Fatigue Properties of Metals," unpublished draft submitted to the Fatigue Design Committee of SAE ISTC Division 4, April, 1964.
50. Morrow, JoDean, Halford, G. R., and Millan, J. F., "Optimum Hardness for Maximum Fatigue Strength of Steel," *Proceedings,* First International Conference on Fracture, Sendai, Japan, 1965, Vol. 3, pp. 1611–1635.
51. Morrow, JoDean, and Sinclair, G. M., "Cycle-Dependent Stress Relaxation," ASTM STP 237, American Society for Testing and Materials, 1958, pp. 83–109.
52. Nair, K. D., and LeMay, I., in *Electron Microfractography,* ASTM STP 453, American Society for Testing and Materials, 1969, pp. 134–146.
53. Neuber, H., *Theory of Notch Stresses: Principles for Exact Stress Calculation,* trans. from the German by F. A. Raven for the David Taylor Model Basin, U.S. Navy, 1945. Also published by Edwards Brothers, Ann Arbor, Michigan, 1946.
54. Ohji, K., Miller, W. R., and Marin, J., "Cumulative Damage and Effect of Mean Strain in Low-Cycle Fatigue of a 2024-T351 Aluminum Alloy," *Journal of Basic Engineering* (Series D of *Transactions of the ASME*) Vol. 88, Dec. 1966, pp. 801–810.
55. Osgood, C. C., *Fatigue Design,* Wiley-Interscience, New York, 1970.
56. Palmgren, A., "Die Lebensdauer von Kugellagern," *Zeitschrift des Vereins Deutscher Ingenieure,* Vol. 68, 1924.

57. Paris, P. C., and Erdogan, F., "A Critical Analysis of Crack Propagation Laws," *Journal of Basic Engineering* (Series D of *Transactions of the ASME*) Vol. 85, Dec. 1963, pp. 528–534.
58. Peterson, R. E., *Stress Concentration Design Factors,* John Wiley & Sons, New York, 1953.
59. Raske, D. T., and Morrow, JoDean, in *Manual on Low Cycle Fatigue Testing,* ASTM STP 465, American Society for Testing and Materials, 1969, pp. 1–26.
60. Raymond, M. H., and Coffin, L. F., Jr., "Geometrical Effects in Strain Cycled Aluminum," *Journal of Basic Engineering* (Series D of *Transactions of the ASME*) Vol. 85, 1963, pp. 548–554.
61. Rice, J. R., in *Fatigue Crack Propagation,* ASTM STP 415, American Society for Testing and Materials, 1967, pp. 247–311.
62. Richard, T. G., and Sandor, B. I., "The Cyclic Stress-Strain Response of High Strength Cast Iron," *Proceedings,* International Conference on Mechanical Behavior of Materials, Kyoto, Japan, 1971, in press.
63. Richards, C. W., *Engineering Materials Science,* Wadsworth Publishing Co., Belmont, Calif., 1961.
64. Sachs, G., in *Fracturing of Metals,* American Society for Metals, Cleveland, 1948.
65. Sandor, B. I., in *Achievement of High Fatigue Resistance in Metals and Alloys,* ASTM STP 467, American Society for Testing and Materials, 1970, pp. 254–275.
66. Sandor, B. I., and Morrow, JoDean, "Alleviation of Fatigue Damage," *Proceedings,* Third Conference on Dimensioning and Strength Calculations, Budapest, 1968, pp. 215–225.
67. Schijve, J., in *Fatigue Tests of Aircraft Structures: Low-Cycle, Full-Scale, and Helicopters,* ASTM STP 338, American Society for Testing and Materials, 1963, pp. 193–215.
68. Serensen, S. V., and Shneiderovitch, R. M., "Design of Construction Elements Under Low-Cycle Loading," *Proceedings,* Third Conference on Dimensioning and Strength Calculations, Budapest, 1968, pp. 227–234.
69. Sinclair, G. M., and Craig, W. J., "Influence of Grain Size on Work Hardening and Fatigue Characteristics of Alpha Brass," *Transactions Quarterly,* American Society for Metals, Vol. 44, 1952, pp. 929–948.
70. Smith, C. R., *Tips on Fatigue,* NAVWEPS 00-25-559, U.S. Government Printing Office, Washington, D.C., 1963.
71. Swanson, S. R., "Random Load Fatigue Testing: A State of

the Art Survey," *Materials Research and Standards,* Vol. 8, No. 4, April 1968.

72. Tetelman, A. S., and McEvily, A. J., Jr., *Fracture of Structural Materials,* John Wiley & Sons, New York, 1967.
73. Thompson, N., and Wadsworth, N. J., "Metal Fatigue," *Advances in Physics* (Quarterly supp., *Philosophical Magazine*), Vol. 7, No. 25, Jan. 1958, pp. 72–169.
74. Topper, T. H., and Biggs, W. D., "The Cyclic Behavior of Mild Steel at Low Endurances," *Applied Materials Research,* Vol. 5, No. 3, July 1966, pp. 131–137.
75. Topper, T. H., and Sandor, B. I., in *Effects of Environment and Complex Load History on Fatigue Life,* ASTM STP 462, American Society for Testing and Materials, 1970, pp. 93–104.
76. Topper, T. H., Sandor, B. I., and Morrow, JoDean, "Cumulative Fatigue Damage Under Cyclic Strain Control, *Journal of Materials,* Vol. 4, No. 1, March 1969, pp. 189–199.
77. Topper, T. H., Wetzel, R. M., and Morrow, JoDean, "Neuber's Rule Applied to Fatigue of Notched Specimens," *Journal of Materials,* Vol. 4, No. 1, March 1969, pp. 200–209.
78. Tucker, L. E., and Olberts, D. R., "Fatigue Properties of Gray Cast Iron," Paper No. 690471, Mid-Year Meeting, Society of Automotive Engineers, Chicago, May 1969.
79. Valluri, S. R., "Some Recent Developments at GALCIT Concerning a Theory of Metal Fatigue," *Acta Metallurgica,* Vol. 11, 1963, pp. 759–775.
80. Vitovec, F. H., "Effect of Relaxation on the Behavior of Materials Under Combined Alternating and Static Stress," *Journal of Basic Engineering* (Series D of *Transactions of the ASME*) Vol. 82, June 1960, pp. 441–446.
81. Weiss, V., Sessler, J., and Packman, P., "Effect of Several Parameters on Low Cycle Fatigue Behavior," *Acta Metallurgica,* Vol. 11, July 1963, pp. 809–816.
82. Wetzel, R. M., "Smooth Specimen Simulation of the Fatigue Behavior of Notches," *Journal of Materials,* Vol. 3, No. 3, Sept. 1968, pp. 646–657.
83. Whiteson, B. V., Phillips, A., Kerlins, V., and Rawe, R. A., in *Electron Fractography,* ASTM STP 436, American Society for Testing and Materials, 1968, pp. 151–178.
84. Zackay, V. F., Parker, E. R., Fahr, D., and Busch, R., "The Enhancement of Ductility in High-Strength Steels," *Transactions Quarterly,* American Society for Metals, Vol. 60, No. 1, March 1967, pp. 252–259.

Index

Amplitude: of stress or strain, 18; not constant, 64-84 *passim*, 143

Basquin's exponent, 47
Bauschinger effect, 12
Beachmarks, 142-43
Buckling, 98-100

Coffin-Manson relation, 46
Composite materials: model of, 121; possible hysteresis loop for, 132-33; mean stress relaxation in, 134; by electroplating, 138-39; for optimum performance, 150-51
Creep: cyclic, 98-103, 133-34; strains in distorted hysteresis loop, 132-33; enhanced by cyclic deformations, 138

Damage accumulation: theories of, 68-69; limiting assumptions in use of theories of, 70, 84; after prestrains, 90-92; impossibility of, in case of fully repetitive behavior, 139-40; retardation of, 152-54
Dislocations: and plastic deformations, 22; in strain-aging, 136
Ductility: and strength, 16, 149-51; exhaustion of, 48-49, 90-92

Elastic behavior in fatigue, 19, 139-40
Elastic strain: monotonic, 8; recovery of, 8-11; cyclic, 22-23, 35-36; vs. life, 43, 45-46, 52

Fatigue ductility: exponent, 44-45; coefficient, 44-45, 47, 61-63; correlation of, with monotonic ductility, 47; meaning of exponent, 48-50, 154
Fatigue limit: definition of, 39; discussion of, 74-81 *passim;* hypothetical example of, 139-40
Fatigue strength: definition of, 39; coefficient, 44, 46, 62-63; exponent, 44, 46-47; reduction factor, 114

Fracture: resistance to, 14; Stages I and II of, 50; criterion, 55-56; appearance at different lives, 97; under mean stresses, 103, 134; prediction of, in notched members, 120; under compressive loading, 137-38; in electroplated parts, 138-39; appearance of, 141-43; origin of, 142; detection of, 145-46; retardation of, 152-54. *See also* Toughness

Goodman diagram, 94-95
Grain size, 151-52

Hardening: under stress or strain control, 26-37 *passim*, 78-81; in step loading, 66-67; and cyclic creep, 99-100; and mean stress relaxation, 108-9; in notched member, 122
Hysteresis loop: definition of, 21; notation of, 22; energy associated with, 23; cycle-dependent changes in, 28-31; recommendation for recording of, 53; based on plastic strain, 58; creeping or walking of, 98-99; in mean stress relaxation, 107-9; distorted, 132-33; for relatively tough material, 150

Inelastic behavior in fatigue. *See* Creep; Hysteresis loop; Plastic strain
Instability: complex metallurgical, 134-35. *See also* Buckling; Creep; Softening
Instrumentation: limitations of, 24, 155; extensometer, 25; view of equipment, 42; mentioned, 20

Linear damage theory: definition of, 68-69; limitations of, 70, 91, 112; on the basis of plastic strains, 71-84
Low-cycle fatigue: life range of, 40, 51; importance of, 41

Manson-Coffin relation, 46

Mean strain: definition of, 18; reduction of life by, 90–91.
Mean stress: definition of, 18; in step loading, 65, 70; and material properties, 95–96; effect of small values of, 97; effective, under fully reversed loads, 133; relaxation of, in inhomogeneous material, 134
Micromechanisms: in crack initiation and propagation, 50, 154; of damage under mean stresses, 89; of damage rates after prestrains, 91; of mean stress effects, 100–103; of strain-aging, 136; by hypothetical example, 139–40; by fracture surface observations, 141–43
Miner's rule, 68–69, 112
Model of: elastic and plastic deformations, 10–11; cyclic plastic deformations, 21–22; exhaustion of ductility, 48; random loading, 64–65; surface peening, 104–5; notched member's behavior, 121; composite material's behavior, 121, 150–51

Neuber's approach to the notch problem, 118

Optimization, 146–54
Overloads; in notched member, 124; beneficial or harmful, 126; practical significance of, 127–30

Palmgren-Miner law, 68–69, 112
Plastic strain: monotonic, 6–11; definition of fracture ductility, 14–15; cyclic, 22–23, 35–36; vs. life, 43–46, 49, 52; and transition life, 51; very small levels of, 75–82 *passim,* 155; in small critical regions, 89, 155; and mean stress relaxation, 109–11; concentration factor, 117; reduction of, by strain-aging, 136; significance of, by hypothetical example, 139–40
Plastic zone: size of, 57
Prestrain: monotonic, 90–91; in design considerations, 91, 127–30; cyclic, 91–92
Programmed loading. *See* Amplitude, not constant

Random loading. *See* Amplitude, not constant
Relaxation of stresses, 102–3, 106–11, 126–27, 132, 134. *See also* Softening
Reversals: in loading, 11–13; to failure, 61–63

Sequence effect in step loading, 65, 70
Size effect, 96–97, 137
S–N curve: definition of, 39; inadequacy of, 40; related to strain vs. life curve, 52
Spectrum loading: crack advance during, 143. *See also* Amplitude, not constant
Softening: under stress or strain control, 26–38 *passim,* 78–81; in step loading, 66–67; and cyclic creep, 98–99; and mean stress relaxation, 107–8; in notched member, 122
Stabilization: of cyclic stress-strain curve, 38; of mean stresses, 109–10, 134. *See also* Creep; Hardening; Softening
Stages I and II in cracking, 50
Strain: sign convention for, 12–13; range, 18
Strain-hardening exponent: monotonic, 8, 35, 37, 59–63; cyclic, 35–37, 61–63
Strength: and ductility, 16, 149–51; increase of, by strain-aging, 136, 154; changes in, by electroplating, 138–39
Stress: sign convention for, 12–13; range, 18; concentrations, 100–103, 114, 116–17, 138–39, 143–44, 151, 153
Striations on fracture surface, 143

Temperature: effect of, 135–36
Test equipment: view of, 41; for damage detection, 145–46. *See also* Instrumentation
Total strain: monotonic, 7–11; components of, 8; cyclic, 22–23, 35–36; vs. life, 43, 45–46, 52; concentration factor, 117
Toughness: definition of, 14–15; of ideal material, 15–16; in fatigue, 54–55, 58–59, 149–51

Ultimate strength: changes in, 37

Volume effects. *See* Composite materials; Size effect

Work. *See* Toughness

X–Y plotting. *See* Instrumentation

Yielding. *See* Plastic strain

DESIGNED BY GARY GORE

COMPOSED BY EDWARDS BROTHERS, INC., ANN ARBOR, MICHIGAN

PRINTED AND BOUND BY CUSHING-MALLOY, INC., ANN ARBOR, MICHIGAN

TEXT AND DISPLAY LINES ARE SET IN *LINOTRON* TIMES ROMAN

Library of Congress Cataloging in Publication Data
Sandor, Bela Imre.
Fundamentals of cyclic stress and strain.
Bibliography: p. 159-64
1. Materials—Fatigue. 2. Strains and stresses.
I. Title. II. Title: Cyclic stress and strain.
TA407.S23 620.1'123 70-176415
ISBN 0-299-06100-0